RAT CITY

445

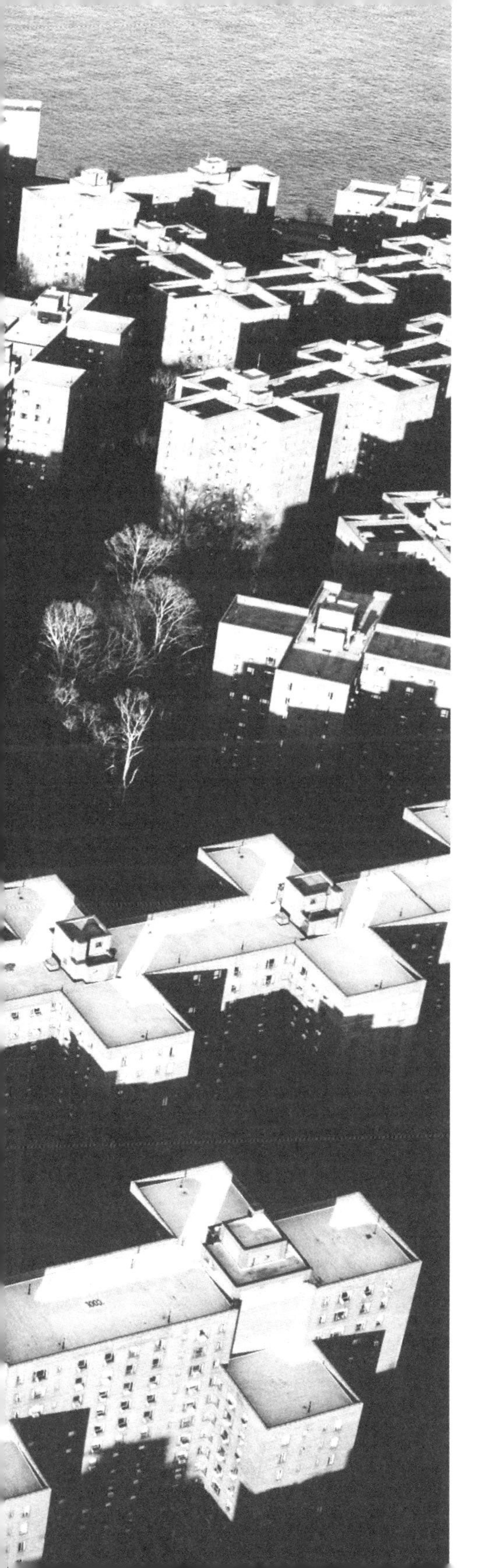

JON ADAMS AND
EDMUND RAMSDEN

RAT CITY

OVERCROWDING AND URBAN DERANGEMENT IN THE RODENT UNIVERSES OF JOHN B. CALHOUN

MELVILLE HOUSE
BROOKLYN · LONDON

RAT CITY: OVERCROWDING AND URBAN DERANGEMENT IN THE RODENT UNIVERSES OF JOHN B. CALHOUN

First published in 2024
by Melville House

First Melville House Printing:
May 2024

Melville House Publishing
46 John Street
Brooklyn, NY 11201
and
Melville House UK
Suite 2000
16/18 Woodford Road
London E7 0HA

mhpbooks.com
@melvillehouse

ISBN: 978-1-68589-099-5
ISBN: 978-1-68589-100-8 (eBook)

Library of Congress Control Number:
2024933885

Designed by Sofia Demopolos

Printed in the United States of America

10 9 8 7 6 5 4 3 2 1

A catalog record for this book is available from the Library of Congress

"Where once were hardly solitary cottages, there are now large cities . . . everywhere are houses, and inhabitants, and settled government, and civilized life. What most frequently meets our view (and occasions complaint), is our teeming population: our numbers are burdensome to the world, which can hardly supply us from its natural elements; our wants grow more and more keen, and our complaints more bitter in all mouths."

TERTULLIAN, C. 200 AD

"If we could solve all the rat problems we would be well on the way to solving many of the critical issues facing mankind today. Perhaps we just might better quit looking directly at ourselves for a while, and instead look at rats to see what that avenue of searching tells us about ourselves."

JOHN B. CALHOUN, 1967

CONTENTS

PART THREE: REVELATIONS

PREFACE

PRIMATE CITIES

This is the house that Jack built.
This is the malt
That lay in the house that Jack built.

This is the rat,
That ate the malt
That lay in the house that Jack built.

ENGLISH NURSERY RHYME, TRAD.

SOME six hundred and ninety thousand people live within the nineteen square miles of the Edogawa ward of Tokyo.

That's a population density of over thirty-six thousand people per square mile. Approximately one in every seventy of those never leaves the house. Nationwide, it is estimated there are over a million and a half people living like this. Nearly all are housed with families, although most do not socialize with them. The family home is a place of purely physical sustenance, and almost total psychic withdrawal.

In Japan, the housebound are called the *hikikomori*. Shut-ins. Isolates. Practitioners of a modern and distinctly urban form of eremetic monasticism. No one is quite sure what causes this, or why it is happening now. But the phenomenon seems to be spreading, with similar cases reported in other Asian countries and in the West, where the trope of young men holed up in their parents' basements has become a chat-board cliché. The Japanese population is shrinking, with birth rates falling as fewer and fewer young people form relationships. More and more hikikomori are reported each year.

John B. Calhoun died in 1995, but he would have recognized the hikikomori. Complete social withdrawal was a typical response to overcrowding in his research, one of a suite of behaviors that emerged under conditions of elevated population density. His term for this particular reaction was "social autism." Other typical responses included aggression, gang formation, and obsessive communal eating and drinking.

Jack Calhoun, though, had not been studying human societies, but colonies of rats.

He had first begun his work with crowded rodents in the 1940s, fencing off a clearing in the woods behind his house. By the 1970s, his rats were living in large, elaborate enclosures within the animal houses of the National Institute of Mental Health, better known by its acronym: NIMH. Calhoun had lobbied long and hard to have special facilities in which to conduct his experiments. His colleagues affectionately called his lab "the house that Jack built."

Inside were the rat cities of NIMH. Calhoun called them "universes" because they were self-contained environments, worlds unto themselves, where all of the somatic needs of the rats were met. Unlike a laboratory cage, which is designed to pacify and contain the occupant by isolating it from others, Calhoun carefully designed the enclosures to allow his animals opportunity to exhibit their natural social behaviors. They were designed to create "rat utopia," and, just as in the wild, the rats formed colonies and raised families. They had abundant food, water, and bedding, plenty of nest boxes. And, until the populations reached a certain size, they had enough space.

But when population density increased beyond a certain point, the social fabric began to unravel.

In the second and third generations, as the rats' universe became increasingly crowded, abnormal behaviors began to emerge in the rat cities. Fertility rates dropped; infant neglect rose. The usual mating rituals were abandoned and replaced by aggressive and persistent nonreproductive sexual activity. Homosexuality became commonplace. Violence, which among animals of the same species is nearly always ritualistic and seldom causes any serious damage to combatants, became endemic. Rats in the wild merely nip one another, but in Calhoun's animal houses, those rules ceased to hold. The rats of NIMH deployed the full force of their jaws, using slashing motions to penetrate through skin and muscle, leading to disembowelment and tail amputation. Private burrows were invaded, and the young nested within were cannibalized. Rat utopia became hell.

Complete social withdrawal became a survival strategy. The isolates, Calhoun's hikikomori, were sleek and glossy, well fed, calm, docile. They often stayed high above the floor of the pens, avoiding the violence and sexual activity rampant below. Calhoun had a name for them: *the Beautiful Ones*. From an evolutionary perspective, their existence was baffling: "For their whole life span they failed to reproduce," he marveled. "They might as well have died at birth."

Eventually, social disruption was so severe that no new young survived past infancy and reproduction ceased altogether. The colonies depopulated toward zero. The trauma was intergenerational: even when individuals born in the later stages of the experiment were removed from the universe and placed in more comfortable surroundings, they failed to develop any of the normal behaviors and produced no viable offspring. From 1947 until 1983, Jack repeated variations on his experiments, tweaking variables, trying to find a strategy by which the rats could survive overcrowding. Each iteration terminated in colonial extinction, the sterile and hermetic Beautiful Ones, asocial hikikomori, always the last survivors.

Jack Calhoun lived through a period of massive population growth. At his birth, in 1917, the global population was a little under two billion. By his death, it was almost six billion. Today, it is in excess of eight billion. Demographers studying that trend discovered that the human population was growing exponentially, each doubling occurring in half the time it had taken for the previous doubling. In 1960, one scientist playfully calculated that, if this continued, the human population would approach infinity on Friday the thirteenth of November, 2026.

Calhoun was trying to find a way to solve the problem of crowding. He didn't believe it was possible to halt population growth—or rather, he felt that the possible routes were impossibly undesirable. Through studying rodents, he wanted to find out if and how humans might cope when social density in the cities approached the densities in his universes.

WHEN NEWS of Calhoun's work first emerged in the early 1960s, it was into a culture undergoing huge social upheaval. Inner-city rioting, motiveless serial murder, sexual deviancy. As an explanation, crowding seemed to fit the data. Colloquially, we are well aware that crowded places are stressful. It's entirely familiar today to say something like "I get stressed when people intrude on my personal space" or "crowded places get my adrenaline going." But actually, these are all quite recent ideas. The link between adrenaline and stress was only discovered in the 1910s, the term "stress" itself only arrived in the 1930s, and "personal space" wasn't a thing until the late 1950s. "Stress" and "personal space" and "the fight-or-flight response" were concepts that emerged from researchers who, like Calhoun, were studying how humans and other animals negotiate social relations and defend territory.

Some of them were architects, or anthropologists, or psychiatrists. Many were ecologists. Nearly all used animal studies to inform and explain their work. In applying what they had learned about animal behavior to the study of human society, startling continuities became apparent. Stress caused by proximity to others caused all manner of problems. When Jean-Paul Sartre wrote "*L'enfer, c'est les autres*"—hell is other people—it was in the context of a play where the characters were unable to leave the room.

Calhoun's work helped people think about how the design and layout of a room might change the experience of people within it. Psychiatrists studying the behavior of mental patients on open wards realized that in addition to whatever else was wrong with them, the patients were also stressed by unwanted and constant social contact. They felt vulnerable, exposed. Simply by putting up screens to break up lines of sight in large spaces, or rearranging the chairs so people didn't have to face one another, they were able to make their patients feel more comfortable.

We all put up screens. A wall is a screen, or a closed door. On a crowded subway train, a newspaper is a screen, or a book. A screen is a screen, too. The electronic devices we carry block the need for social engagement, reduce the risk of eye contact with strangers. In crowded places, we retreat behind them, a black mirror held up to the world without. Such distractions are necessary: the city can be a uniquely uncomfortable habitat in which to live.

The rat cities of NIMH were models of our own cities: places in which unwanted contact with strangers forced withdrawal or confrontation, where the pursuit of privacy would vie with the need to connect. The global trend toward urbanization over the last century and a half has seen an ever-greater proportion of an ever-larger global population living in ever-denser agglomerations. A physicist in the 1940s watching the seemingly inexorable concentration of people drawn into the urban core described the phenomenon as "demographic gravitation."

In most countries, city size follows a curious power law, whereby the largest city is twice as populous as the second largest city, and the third largest city a third as populous, and so on. But in some countries, there is one city much larger than any others. Dublin is five times more populous than Cork; Kabul is more than eight times the size of Kandahar; while in Ethiopia, Addis Ababa is almost ten times larger than Adama. Phnom Penh is fifteen times larger than Cambodia's second city of Siem Reap. Almost ten percent of the population of Japan lives in Tokyo, where the hikikomori wait out their quiet lives. Geographers have an odd term for such places. They call them "primate cities." There are relatively few primate cities.

But every city is a rat city. Wherever there are people, there will be rats. Like us, and because of us, they have become a global species. Of all the continents, only Antarctica has no rats. But it also has no cities. The largest inhabited area that has managed to eradicate rats is the sparsely populated Canadian province of Alberta, and only by dint of vigilant rodent control, abetted by sufficient peripheral wilderness. New Zealand hopes to replicate this feat by 2050.

It was rodent eradication that first drew Jack Calhoun from rural Tennessee to inner-city Baltimore, where he was enlisted to help combat a wartime rat infestation. His story is the spindle around which all else is organized in this book. But telling his story requires following the rat from the baiting pits of Victorian London to the pristine laboratories of Johns Hopkins and on to the rat cities of NIMH. It involves the emergence of ecology and ethology, and how these new sciences led to new ways of thinking about the place of humans both in relation to other animals and to the habitats we have built for ourselves.

Habitat matters. Darwin's theory of evolution established that the physical form of an organism reflected its adaptation to a particular habitat. Calhoun came from a school of thought that emphasized how behavior, too, was adaptive, and that what sorts of things an animal did—whether it gathered in packs or lived alone, whether it built nests or dug burrows—were also adaptations to particular habitats. And so, while his colleagues performed experiments on animals kept in standardized cages, Calhoun experimented with different cages and left the animals themselves alone. What sort of box you kept an animal in would affect what it did.

In the early 1970s, Jack Calhoun was briefly famous. His work was regularly featured in the press; he was consulted by the United Nations, introduced to the Pope, and even put forward for the Nobel Prize. By the 1980s, as the focus of mental health shifted away from environmental causes and toward pharmacological fixes, his scientific stock value began to decline. Meanwhile, a growing queasiness over the social and political ramifications of "population control" meant that the threat of overpopulation largely disappeared from polite discourse, and with it, Calhoun's relevance.

In recent years, as a new generation learns of the striking parallels between his crowded rodents and our urban predicament, there has been renewed interest in the rat cities of NIMH. Much of this is focused on "Universe 25," one of Calhoun's later experiments. Visually arresting and viscerally repellant, Universe 25 has become a shorthand for civilizational collapse. But the actual experiment has receded behind a caricature, become a meme. Universe 25 was the most notorious of his projects largely because it was the most photographed. But it was only one of a series of different research cycles, each exploring an issue raised by the previous cycle. And each linked in an intellectual trajectory he pursued assiduously over four decades.

Calhoun was aware of the macabre appeal of his work and sought to use the press attention to try to spread his optimistic and increasingly urgent message that humanity need not endure the same fate as his rodents. That positive message has largely been lost, and to date no full account of his scientific career has been written. This book is an attempt to fill in the history leading up to Universe 25 and beyond. Ours is not an attempt to evaluate the merit of his work, but simply to set the rat cities in their scientific and historical context, and to answer the questions they raise: Why did he create his strange model universes? Why did he use rats? And what did he and others think a pen full of swarming rodents really told us about life in our human cities?

PART ONE
GENESIS

CHAPTER 1

A NEW WORLD

TWO ships, the Ark and the Dove, set sail from England in late November 1633, bound for the Americas. Financed by Cecil Calvert, the second Baron Baltimore, they carried nearly one hundred and fifty souls for settlement in the new colony of Maryland. Several days into the crossing, the much smaller Dove was reported missing in high seas and presumed lost. The Ark sailed on, absent its pinnace, though not without further event. After casks of wine were opened for celebratory drinks on Christmas Day, one passenger would later recall that "it was soe immoderately taken as the next day 30 sickened of fevers, whereof about a dozen died afterward."[1] On January 3, the Ark arrived in Barbados, and two weeks after that so did the Dove, which had taken shelter in Plymouth port and survived the storm. They proceeded in convoy up the Eastern coast to Jamestown, and onward to disembark on what is now St. Clement's Island in the Chesapeake Bay. The date of their arrival is still marked as "Maryland Day," and the city just north of their landing named for their benefactor, and Maryland's first governor, Lord Baltimore. Unlike its biblical namesake, Baltimore's Ark wasn't carrying two of every kind of beast, but it is unlikely it was carrying only one, and therefore intriguing to wonder whether, when the Ark discharged its surviving human cargo of colonists onto American soil, it didn't also deposit a colony of rats.

Rats were not native to the Americas but arrived with the Europeans. At first, they would have been black rats, *Rattus rattus*, which had radiated out from northern India, migrating into Europe about a thousand years ago following caravans along the spice roads, later seeding Medieval Europe with the Black Death and eventually stowing away on vessels bound for the New World. By the eighteenth century, the larger Norway or brown rat, *Rattus norvegicus*, had almost wholly displaced the black rat in most of Europe, and upon arrival in North America sometime in the 1750s proceeded also to dominate there.

Baltimore's first Norway rats might have found a town of only a few hundred settlers, but the city grew rapidly. When it was incorporated in 1797, the population stood at around six thousand. By 1820, when English surveyor Thomas Poppleton was hired to map and contain the sprawl, that number had increased tenfold to well over sixty thousand.[2] The map was also a plan—a trellis on which the city might grow. Poppleton produced the now-familiar orthogonal hierarchy of main streets feeding side streets, forming rectangular city blocks threaded with narrow service alleys. Filling up and soon overspilling Poppleton's grid, Baltimore became a major trading port and industrial hub, with mills, shipyards, and factories appearing all along the shores of the Potomac. Great wealth was accrued, and millionaire philanthropists competed to bestow ever-grander symbols of their success upon the city: the Enoch Pratt Free Library, the Walters Art Gallery, and, in 1876, Johns Hopkins University and Hospital, America's first research-focused institute of higher education. By 1900, there were half a million residents, including many freed slaves from the south.

Inside the grid of streets, terraces of row housing provided affordable and efficient accommodation to the endless stream of workers. Fitting as many as seventy properties per block, the row-house design maximized the available space with the minimum building materials. But that density embedded a precarity, and after an explosion at a dry goods store in 1904, fire travelled easily through the terraces, eventually laying waste to one hundred and forty acres of downtown Baltimore. Rebuilding was swift, but as the influx of migrants both from overseas and from the south continued, the population rapidly outstripped housing supply; meanwhile, zoning laws were soon to be introduced, preventing black people from moving into white neighborhoods. The trellis became a trap, rigidly containing populations within their allotted cells.

So Baltimore followed a trajectory common to many American cities: rapid expansion in the nineteenth century, an influx of freed slaves following the Civil War, industrial boom, economic collapse, political corruption, the obligatory Great Fire, and Jim Crow gerrymandering embedding structural inequalities that would entrench a seemingly indelible legacy of racial disparity and social division. All laid out over a regimented grid of city blocks that exerted a subtle but very real form of control over the inhabitants they housed and divided.

And yet, for every way that Baltimore was typical, it was peculiar. During the Civil War, Baltimore had been a "middle city," poised between North and South, and was embedded now within the increasingly connected, increasingly dense megalopolis that stretched from New York through Philadelphia to the capital in DC. When Johns Hopkins was inaugurated in 1876, its first president, Daniel Coit Gilman, had insisted on the importance of place. The university was to sit "in the heart of the city, accessible to all." Rather than the ivy-coated gothic colleges of Harvard and Yale, it was to occupy "a series of modern institutions; not a monumental, but a serviceable group of structures." After all, as Gilman put it: "The middle ages have not built any cloisters for us; why should we build for the middle ages? In these days laboratories are demanded on a scale and in a variety hitherto unknown." Forward-looking, focused on research and learning, Gilman said the "object of the university is to develop character—to make men."[3] And hence Johns Hopkins, the university and hospital founded by Quakers, beholden to "build men, not buildings," resisted the opportunity to relocate, remaining downtown even as the city thickened and deteriorated around it. That physical immersion among the urban poor would come to shape both the questions its researchers asked and the evidential materials available, while the twinned functions of clinical hospital and academic university made fluid the transfer of information between public health policy and theoretical research: informing the former, grounding the latter.

Following a pronounced economic slump and mass unemployment during the Great Depression, the outbreak of World War II renewed Baltimore's fortunes. The shipyards and carmakers converted their production lines and began churning out materiel for the war effort at an astonishing rate: almost four hundred Liberty ships for the navy, over five thousand B-26 bombers for the air force.[4] Great flows of workers entered the city, but with little new housing stock, many row houses were hastily converted for multiple occupancy. Amid the increased population of humans, an increasing population of rats was becoming a problem.

"It was as if they had taken over the city," an intern at Johns Hopkins Hospital would later recall. "They scuttled about: aggressive, fearless, upsetting garbage cans and fighting ferociously with each other."[5] Reports began to accumulate of infested warehouses, and over a hundred cases of rat bites—mostly suffered by infants.[6] With the city such a crucial part of the war effort and resources already stretched, any contamination of food stores or damage to infrastructure and apparatus by vermin was intolerable. Not only that, but there was concern that rats, which had long been a natural vector of disease, might be actively weaponized by the Axis powers to spread bubonic plague on US soil. And so it came to pass that in 1942, while the world was at war, the City of Baltimore opened a new battlefront: they declared war on the rat.

YELLOW FEVER ravaged Philadelphia during the fall of 1793. Headaches, nausea, vomiting, and muscle pain gave way to liver failure, causing the distinctive yellowing of the skin and subsequent bleeding from the mouth, nose, and eyes. When the first cases had been detected in August, Dr. Benjamin Rush, signatory to the Declaration of Independence and later memorialized as "the Father of American Psychiatry," sought, unsuccessfully, to treat the sick with bloodletting and mercury chloride. By November, some five thousand citizens, ten percent of the city's population, were dead. Panic spread along the Atlantic seaboard. Fearing a similar decimation, quarantines against goods and persons were introduced in Baltimore, Boston, and New York. Thomas Jefferson thought the epidemic providential, a corrective to American urbanization. He would later write to Rush: "The yellow fever will discourage the growth of great cities in our nation; & I view great cities as pestilential to the morals, the health and the liberties of man."[7] Baltimore, unwilling to see its growth discouraged, appointed its first health officers that same year and promptly established a permanent Board of Health. Other cities followed. When the Boston Board of Health was set up seven years later, its first president was Paul Revere.

By the 1940s, the Baltimore Board of Health was celebrating its fifteenth decade of continuous operation. And yet, successive attempts to improve living conditions within the municipality had achieved little, and sometimes made things much worse. Long before the influx of war workers, much of the city's housing was already in a dolorous state. An article from 1892 casts a racially prejudicial eye over the conditions in Baltimore's Pigtown district:

> Open drains, great lots filled with high weeds, ashes and garbage accumulated in the alleyways, cellars filled with filthy black water, houses that are total strangers

> to the touch of whitewash or scrubbing brush, human bodies that have been strangers for months to soap and water, villainous looking negroes who loiter and sleep around the street corners and never work; vile and vicious women, with but a smock to cover their black nakedness, lounging in the doorways or squatting on the steps, hurling foul epithets at every passerby; foul streets, foul people, in foul tenements filled with foul air; that's "Pigtown."[8]

Fifteen years later, in 1907, social reformer Janet E. Kemp published an influential report, *Housing Conditions in Baltimore*, which aimed to improve "sanitary surroundings" for the poor. She had begun her survey in 1903, but work was interrupted by the Great Fire of 1904. When she resumed in 1906, she might have wished the fire had claimed even more buildings. Documenting the open sewers, makeshift abattoirs, flooded basements, overcrowded lots, and overflowing hogshead barrels serving as shared toilets, Kemp's report makes for grim reading.

Housing Conditions In Baltimore is significant because Kemp identified a flaw at the heart of unregulated capitalism: "Chief among the causes which operate to bring about such conditions may be the mentioned the landlord's desire to increase his profits." A landlord with a plot of land will attempt to extract as much rent as possible, which led to overcrowding and minimal maintenance. The consequences were predictable:

> This may, indeed, mean the slow destruction of the tenant, but it is safe to say that we are still too far removed from the millennium to hope to control, to any large extent, save by legislation, an evil which has its root in the not unnatural or uncommon desire to secure the largest possible returns on one's investment.[9]

With demand for housing far outstripping supply, the usual market forces failed to kick in to drive standards up, and—because there was no incentive for landlords to improve the buildings—would actually drive them down. Making improvements cut into the profit margin.

Illustrated with photographs of windowless and filthy rooms visible to the photographic plate only by means of a phosphorescent flash, *Housing Conditions in Baltimore* caused a stir of civic embarrassment as much as human sympathy. In the short term, it had the effect of improving situations for the whites—mainly immigrant Poles, Russians, and Germans—but did little to touch the "alleys" where the majority of the city's black population lived. Compounding this was the inevitable racial hierarchy. A mere two generations out of slavery, the black population was generally held to be deficient in character, and therefore blamed for their living conditions in a way the poor whites were not. Kemp's call for remedial legislation would be answered, but disastrously so, as the tacit segregation that already existed was soon to become official.

EVEN AS the Board of Health sought to drive rats out from the city, scientists at Johns Hopkins would soon be importing a different kind of rat back in—one that had been selectively bred for precise laboratory work. For while Kemp was campaigning to sanitize the slums, the rat itself was undergoing a process of "sanitization," and a new role was emerging from the mission to eradicate them, a parallel history that would eventually result in an unlikely convergence between ratcatchers and city reformers.

Ratcatchers had been around for centuries, sending wiry terriers scurrying into barns and basements. In Old World mythology, there is the story of the infestation of Hamelin, a medieval town in Lower Saxony. There, an itinerant ratcatcher had enchanted the rats from their hiding places by playing a tune on his pipe, luring them in a massed hypnotic swarm to drown in the Weser river. When the mayor refused to pay the ratcatcher's fee, the Piper returned and enticed all the town's children away by the same means.

By the early nineteenth century, ratcatchers had developed a secondary market for their harvest. The rats they trapped were held captive, and as many as two hundred at a time emptied into a wood-paneled pit. A terrier would be dropped among them, while punters placed bets on how long it would take the dog to dispatch every last one. Rat-baiting became a popular variation on an old blood sport, but unlike the bears and bulls that had once been baited, rats were cheap and in near-limitless supply. Every once in a while, an albino mutation would be caught. Rather than being thrown into the baiting arena, the white rats were separated out and saved. Ever alert to sources of additional income, the ratcatchers discovered they could sell unusual variants to breeders supplying the burgeoning hobby of rat fancying. And so it came to pass that pampered fancy rats were displayed at pet shows even as their ilk were being tossed to the dogs.

Having fortuitously escaped the baiting pits, white rats not selected for showing were soon to find employ in another of the rapidly proliferating institutions of the nineteenth century: the scientific laboratories. The first albino rats were used for anatomical dissection in Paris in 1850,[10] and they quickly became a popular choice for scientists across Europe.

Visiting the University of Chicago in the early 1890s, a Swiss neuropathologist called Adolf Meyer brought the white rat to America. Meyer was using the albino rat as a teaching aid in his lectures on the anatomy of the nervous system. Neurologist Henry Hubert Donaldson was impressed and secured a temporary post at Chicago for Meyer, and together they began using rats to investigate the growth and development of the brain. Meyer subsequently found more permanent work at a psychiatric hospital in Kansas, initially as a pathologist but gradually migrating into clinical psychiatry. Attempting to steer psychiatry toward a biomedical model, Meyer and his white rats eventually settled at Johns Hopkins in 1909, where he was to oversee the establishment of the Henry Phipps Psychiatric Clinic. Donaldson, meanwhile, was offered a post at the Wistar Institute in Philadelphia, where he was to lead the neurology division. When Donaldson moved to Philadelphia, he also brought his white rats with him and began a rigorous breeding program. As the rat population at Wistar grew, Donaldson began to hanker after precision. The lab rats certainly looked similar enough, but for research purposes, Donaldson now wondered just how similar it would be possible to make them.

MEANWHILE, BY 1910, Baltimore's seventeenth ward in the city's northwest had become increasingly squalid, overcrowded, and almost wholly black. As those black families who could afford to do so moved out, the boundary of the so-called "Negro district" crept west into the fifteenth and sixteenth ward, and north into the fourteenth. The wealthier white residents called it "the black sea," smeared tar on the white marble front steps of black-owned houses, and attacked a black family who had moved to more prosperous Stricker Street. In July, Mayor J. Barry Mahool was petitioned to "take some measures to restrain the colored people from locating in a white community, and proscribe a limit beyond which is shall be unlawful for them to go."[11]

In response, the City introduced one of its most notorious policies: the Segregated Housing Ordinance. This decreed that black residents must only live within certain zones and would be prevented from moving to other areas of the city. Consonant with the City's longstanding stance on public health, the putative rationale was to reduce the spread of contagion: the especially poor living conditions endured by Baltimore's black population were leading to higher levels of infectious disease. With tuberculosis and typhus rife, containing the most affected areas became a means of containing illness, and racial segregation would now be framed as a public health measure.

There was a yet more sinister dimension. A prevailing belief in social Darwinism—the Victorian idea that saw inequality as the sociological equivalent of survival of the fittest—took the increased mortality rate within the black neighborhoods as evidence that the "weaker races" were, quite naturally, dying out. Sequestered within the ghettos, what was being described as "the negro problem" would eventually solve itself as black populations dwindled toward an inevitable extinction. Here was city planning deployed as assisted genocide. City solicitor Edgar Allen Poe (not the author, but his second cousin twice removed) approved the bill, and on May 15, 1911, Mayor Mahool signed the ordinance into law.

Baltimore was the very first city to introduce such zoning laws, but it would not be last. Surely surprised that Baltimore could get away with overt and, moreover, legalized segregation, Mooresville and Winston-Salem in North Carolina soon followed Baltimore, then Asheville. Over the next two years, a total of twelve American cities, including Atlanta, Birmingham, and St. Louis, would implement similar segregated housing ordinances. Only when the Supreme Court found the practice unconstitutional in 1917 would the spread of what was later called "Apartheid, Baltimore Style" be checked, although other, more insidious methods of segregating housing would continue.

The immediate effect of the racial quarantine was a rapid decline in already dilapidated housing. With their tenants unable to move out, landlords had a quite literally captive market, and black neighborhoods soon became even more crowded, more rundown, more unsanitary. To those who approved of the segregation, here was proof of its necessity. Not only that, but wealthier white citizens, now protected from "the black sea," had even less incentive to improve those regions. When neighborhoods abutting the black ghettos complained of their proximity, the City razed properties inside the perimeter, replacing them with bulwarks of green parkland. The white residents were now better insulated, the black districts just got even smaller. Fully twenty percent of the city's population, Baltimore's black citizens were now crammed into just two percent of its housing stock.

AS BALTIMORE'S elite grappled with racial differences, at Wistar, Donaldson was reflecting on the differences among his rats. If the rat was going to be useful as a tool for exchanging information between different centers of scientific research, it would need to possess a uniformity comparable to the engineers' weights and measures. With the industrial standard of the Sellers screw thread as his model, Donaldson set out to make a standard rat:

> The animal chosen for this purpose should be of the greatest possible uniformity in vitality, weight, structure of its organs and parts, at given ages, and in transmitting power—a stock as nearly homozygous as possible. In other words, a standardized animal is needed for the most accurate results in biological research.[12]

But how could you "standardize" an organism? Every individual specimen had a slightly different genetic makeup, the generational shuffling and recombination of which was both the engine and reserve power of evolutionary adaptation. Mutations and variations were inevitable. Natural genetic variety meant that every pup in every litter would be unique. The answer came from the only female researcher at the Wistar Institute, biologist Helen Dean King. In 1909, at forty years old, King began to selectively inbreed rats from the Wistar colony.

King was well aware of what she called "the almost universal prejudice against inbreeding,"[13] an aversion that wasn't just an ethical qualm but a biological limit. Genetic diversity was crucial for population health as a reservoir of potentially beneficial adaptations. Inbreeding not only reduced the capacity of a population to adapt to any external changes, but also caused recessive traits to build up in the genome, leading to cumulatively severe malformations and eventual sterility. No lesser authority than Charles Darwin himself had concluded that "long-continued close interbreeding is injurious" and that destructive mutations were an inevitable result of "close interbreeding carried on for too long a time."[14] But King had a hunch: perhaps it had simply never been carried on for long enough.

For six years, King mated brothers with sisters, carefully choosing only those offspring that were healthy and shared the characteristics she wanted to preserve. The theory was that by exclusively mating any healthy siblings, the latent genetic variety within the animals could be gradually bred out, and the genome winnowed down to a stable and uniform template, as near identical as possible between generations. Still, her initial results were not encouraging. She writes:

> In the earlier generations the inbred rats exhibited all of the defects which are popularly supposed to appear in any closely inbred stock. Many females [. . .] were sterile, and those that did breed usually produced only one or two litters which were generally of small size. A considerable proportion of the rats were dwarfed, on stunted in their growth, and many of them developed malformation, particularly deformed teeth.

By the sixth generation, the output of healthy offspring was sufficient that she could yield a thousand rats per breeding cycle. And so eventually, refuting Darwin, King was able to demonstrate that inbreeding—if carefully supervised—did not necessarily result in damaging mutations. On the contrary, under these controlled conditions, deleterious mutations could be eliminated, desirable traits preserved, the "perfect specimen" all but achieved. By 1915, after twenty-eight generations of inbreeding, yielding tens of thousands of offspring, she had produced an organism almost completely homozygous. With so little genetic variation, the Wistar rats would now "breed true," yielding reliably identical albino offspring. What King had produced amounted to a race of incest clones.[15]

The uniformity of the Wistar rats made them ideal for scientific experimentation: where genetic diversity was beneficial for wild populations, genetic homogeneity was ideal for scientific study. So invariant were the Wistar rats that Donaldson was able to produce a guidebook for their use: *The Rat: Reference Tables and Data* lays out the exact size and weight of the Wistar rats at different stages of their lives, allowing experimenters to confidently ascribe any deviation from these figures solely to their intervention. Packed in boxes stamped with the Wistar logotype, they were shipped to laboratories across the States. The lab rat became a branded product, and the Wistar would be joined by others: the Long-Evans, the Sprague-Dawley and the Osborne-Mendel, the Brattleboro, the Lewis, and the Royal College of Surgeons rat. Nearly all the variations were crosses with the Wistar stock.

The lab rat became a working model of the human body in miniature, a precision instrument with regularized and known specifications. Donaldson's reference book even includes tables for comparing brain development in rats and the corresponding brain development in humans: a human at two years old is equivalent to a rat at twenty-six days; a human at nine and half years is equivalent to a rat at one hundred and fifteen days.[16]

But the reference tables said nothing about behavior. The only relevant behavioral characteristic was docility: compared to their wild ancestors, the lab rats were extraordinarily passive and easily tamed. Besides, it was one thing to make analogies between rodent and human bodies, quite another to make such analogies between their behaviors. The use of rats for psychological research would require a fundamental shift in how psychologists thought of behavior, and that change would come about as a result of the work of John Broadus Watson.

While Donaldson had been at Chicago, Watson had been a doctoral student investigating how learning occurs—how a brain is able to store new information, how habits are acquired. Because the problem of how the brain produced behavior straddled physiology and psychology, he had several academic advisors, and Donaldson was helping him with the neurology. Watson had been encouraged by another supervisor to study the brains of dogs, but Donaldson intervened and suggested he use rats instead. Watson's thesis was completed in 1903 and published as a book that same year: *Animal Education: An Experimental Study of the Psychical Development of the White Rat, Correlated with the Growth of its Nervous System.* Confident of the scientific value of the work, Donaldson loaned Watson $350 to pay for the printing.[17] At only twenty-five years old, Watson was the youngest Ph.D. in the university's history.

Animal Education was well received, and Watson promptly offered a faculty position at Chicago. Five years later, Watson accepted a more senior post at Johns Hopkins, where he would continue his work with rats. By steering him away from dogs, Donaldson's intervention unwittingly secured the white rat as the preeminent animal model for psychological research. Through Meyer and Donaldson, the rat had become established as a model for studying the brain, but now Watson wanted to use the rat as a means of studying behavior. To do so, he would need to revolutionize the study of psychology, and the rat would be crucial in effecting that change.

IN BALTIMORE, the type of useful legal intervention that Janet Kemp had wished for would be another three decades coming. Concerned once again with increasing overcrowding and a rise in meningitis, tuberculosis, typhus, Weil's disease, and elevated levels of infant mortality, in March 1941 the city council had passed the Ordinance on the Hygiene of Housing, which set guidelines for cleanliness and gave inspectors the powers to fine any residents failing to meet minimum standards. The ordinance empowered City Sanitation officers to compel landlords and tenants to maintain sanitary conditions. It became known as "the Baltimore Plan."

Instead of entirely demolishing slums, the Baltimore Plan introduced "urban rehabilitation"—a process of repairing and improving existing infrastructure and buildings. Importantly, it placed the onus for making these improvements on the property owners. Centralized rebuilding projects were both costly to the municipal budget and ran against the grain of American market capitalism. Where the invisible hand failed, fines helped nudge reluctant landlords to play along.

Rather than attempt to tackle the city all at once, the Baltimore Plan was rolled out one block at a time. Windows were fixed, toilets installed, drains cleared. The process was achingly slow—only one block was completed in the first year of operation—but the Baltimore Plan represented a quite radical step. Wielding an authority that would impress today's homeowner associations, it made tidying up your property a legal obligation: "Every dwelling and every part thereof shall be kept clean and free from any accumulation of dirt, filth, rubbish, garbage or similar matter, and shall be kept free from vermin or rodent infestation. All yards, lawns, and courts shall be similarly kept clean and free from rodent infestation."[18] And so it was that under the Hygiene of Housing Ordinance, rodent eradication became an article of law.

In wartime Baltimore, the rodent problem was exacerbated by a shortage of poison. The plant used to manufacture the most common poison, red squill, was usually harvested around the southern Mediterranean, but exports were now being blocked by fighting in North Africa. The City authorities wondered if modern science might provide a new approach to eradicating rats and, with the support of the National Research Council at the National Academy of Sciences in Washington, the *Rodent Control Project* was born. They turned to the local Johns Hopkins Medical School, where one of the world's leading animal experimentalists was studying the biochemical basis of rodent behavior.

CHAPTER 2

JOHNS HOPKINS

"If someone were to give me the power to create an animal most useful for all types of studies on problems concerned directly or indirectly with human welfare," Curt Richter once wrote, "I could not possibly improve on the Norway rat." For a scientist, it was simply "one of the world's most valuable animals." He spoke as a man with considerable experience, recollecting toward the end of his career that he had worked with "many different animals, such as cats, dogs, monkeys, sloths, rabbits, beavers, porcupines, honey bears, alligators, and others."[1] Of all the animals he had worked with, Curt Richter thought rats were the best, and that's why he wanted to kill them.

Richter was born in 1894 in Colorado, where his family operated a metalworking factory. After his father died when Curt was only eight, he helped his mother run the business, spending his afternoons disassembling and repairing machines, working with cams and springs and magnets. Later in life, he acquired a reputation for designing ingenious devices to perform laboratory experiments. It would be Richter who eventually discovered that animals have within them a "biological clock." In 1912, he moved to Dresden, Germany—his father's homeland—to train as an engineer. The advent of World War I, and the persistent suspicion that the young American student was a spy, drove him back to the US, where he attended Harvard, drifting between disciplines, taking courses on economics, diplomacy, history, and biology. After being discharged from a brief stint of military service, he settled on psychology and enrolled as a doctoral student at Johns Hopkins, where he was assigned John B. Watson as his supervisor.

When Richter arrived at the Johns Hopkins Experimental Psychology Laboratory in 1919, Watson, the unit's director, was at the peak of his career, one of the most significant figures in the history of psychology who had been pivotal in establishing the study of the mind as a rigorous science. But within a year, Watson would be gone, never to hold an academic position again. And Richter, who would later memorialize himself as "a reluctant rat-catcher," would have a lab full of albino rats all to himself. The rat would be central to both men's careers, and they to the scientific career of the rat.

DURING THE early twentieth century, psychology was enmeshed with psychiatry, and both were short on scientific credibility. This was in part due to the paucity of available empirical data about how the mind worked, but more proximately a result of the still dominant legacy of Sigmund Freud, whose approach to understanding the mind had relied upon the subject's own reports of their thoughts and feelings. Introspection-as-evidence presented a conundrum: the only way to know what was happening in someone's head was to ask them, but people are not always reliable reporters of their own thoughts and motives—they can be deluded, or simply dishonest. How then to observe what was going on in people's minds without relying on their subjective reports of conscious states? Freud's answer was to try to read hidden meanings into what people said, as a literary critic might seek a deeper meaning in a text. He asked people about their dreams, and introduced the now-familiar notion of a "subconscious"—a consciousness below consciousness, accessible only to the trained psychotherapist. The considerable interpretative latitude afforded to the therapist by this method meant that Freud's work seemed to align the study of the mind more with the arts and humanities. Watson sought to set the field firmly among the sciences, to remove all that subjective talk of feelings and thoughts and memories, and create in its place an empirical, objective methodology for doing psychology. His revelation was to jettison the conscious reports altogether: ignore what people claimed they thought or felt and observe only their behavioral outputs. He called this approach *behaviorism*.

Watson would remain studiously indifferent to whatever might be happening inside the minds of people, treating the personality as a black box and limiting his conclusions to inferences derived only from observable behaviors. If a person says they are cold, that means nothing unless there is some physiological correlate: goosebumps, shivers, blue lips. Human consciousness need play no part—indeed, must play no part—in its scientific study. As Watson put it in a paper that became known as "The Behaviorist Manifesto": "To make consciousness, as the human being knows it, the center of reference of all behavior, forces us into a situation similar to that which existed in biology in Darwin's time." If it was to mature as a science, psychology had to stop thinking of human brains as special or different from the brains of other animals. "The behaviorist," Watson declared, "recognizes no dividing line between man and brute."[2] In other words, one would study a human exactly as one might a rat or a dog, without speculative recourse to any inner mental activity. Reflexively, this also meant that inferences about the functioning of minds derived from animal studies should be applicable to the functioning of human minds.

When he arrived at Johns Hopkins in 1908, the thirty-year-old Watson was appointed Professor of Experimental and Comparative Psychology and Director of the Psychological Laboratory. The following year, Adolf Meyer, the Swiss neuropathologist-turned-psychiatrist who had first introduced the white rat to American science, also arrived at Johns Hopkins. Meyer was charged with overseeing the Henry Phipps Psychiatry Clinic, which was still under construction. Eventually, Watson's psychology lab would be subsumed under the administrative supervision of the Phipps clinic, and Meyer would grow increasingly at odds with Watson's behaviorist orientation, which he considered unduly reductive and insufficiently engaged with biology. As Meyer saw it, Watson's behaviorist program aimed to eliminate the study of consciousness altogether in favor of machinelike conditioned reflexes.[3] Both men were using rats from Donaldson's Wistar colony: Meyer for neurological experiments, Watson for behavioral experiments.

At Johns Hopkins, Watson was training his rats to navigate mazes, studying how learning could be understood as an automatic process. When we think of "learning," it is usually in the context of our cultural artefacts and institutions—schools, books, lecture halls, apprenticeships. Watson's question: If a lab rat can learn without all these things, but simply through its environmental interactions, how much of our own learning occurs in a similarly unconscious fashion? Would it be possible to manipulate the human environ ment to facilitate such learning, to influence behavior unwittingly?

Watson's approach was inspired by the Russian scientist Ivan Pavlov, who in 1897 had published the results of his now-iconic experiments with dogs. Those experiments had demonstrated that an organism underwent physiological changes in response to an anticipated event: when Pavlov sounded an alarm (he never used a bell) before every mealtime, the dogs began to salivate and their stomachs released digestive acids. Eventually, this response occurred even if no food was produced. By teaching dogs to associate the buzzer with food, the experiments established that a response (salivation and the release of gastric acid) might be arbitrarily assigned a stimulus (sound). Watson wanted to investigate the extent to which human behavior could be explained and ultimately controlled using the same sort of Pavlovian conditioning.

Regrettably, Watson's insistence on the equivalence between humans and animals would eventually go a little too far. In order to study the development of emotions as conditioned responses, Watson, along with his research assistant, Rosalie Raynor, set out to intentionally induce a phobia in a human infant. The plan was to show a baby something fluffy and delightful (they first used a white lab rat, but later a Santa Claus mask, a fur coat, a rabbit) and, just as the infant saw it, loudly bang a metal bar with a hammer to shock the baby, and thereby create a cognitive association between fur and fear. "Can we," Watson asked, "condition fear of an animal, e.g., a white rat . . . ?"

Watson and Raynor somehow procured a nine-month-old human infant for the experiment. Historians disagree over where and how they acquired the baby (it is generally believed he came from the hospital adjoining the medical school), or who the child's parents were, but Watson called him "Little Albert" and used him as a laboratory subject for several months. Their experiment was a success by its own lights—Little Albert seemingly did develop a phobia of rats—but Watson and Raynor's notes make for uncomfortable reading: "The instant the rat was shown the baby began to cry. Almost instantly he turned sharply to the left, fell over on left side, raised himself on all fours and began to crawl away so rapidly that he was caught with difficulty before reaching the edge of the table."[4] With the results written up, the baby was returned, and Watson never got around to deprogramming Little Albert of his new phobia.

What ought to have been a scandal of experimental ethics was, in the end, overshadowed by a far more prosaic scandal. Shortly before the publication of their "Little Albert" findings, it emerged that Watson and Raynor had been more than merely research partners. Watson's wife, Mary, found their love letters in her husband's bedside cabinet. The extramarital affair might have been ignored as a domestic issue were it not that Mary (née Ickes) was the sister of prominent politician Harold L. Ickes, which made the story noteworthy enough for extracts of the letters to be published in the national newspapers.

This was hardly the sort of press Johns Hopkins wanted, but as it happened, Watson had already been ratted out. He had confided details of his affair with Raynor in writing to Meyer, who promptly passed the letters on to the university authorities, urging that Watson be let go. "Without clean cut and outspoken principles on these matters," Meyer primly complained, "we could not run a coeducational institution, nor would we deserve a position of honor and respect before any kind of public, nor even before ourselves."[5] And to think that, as a behaviorist, Watson could have so easily concealed it all.

Six days after Meyer's delation, Watson accepted an invitation to resign. Having orchestrated Watson's dismissal, Meyer now arranged for the young Curt Richter—who had not yet completed his Ph.D.—to step up as Watson's replacement, passing over far more senior candidates.[6] So it happened that in 1920 Richter would inherit Watson's new laboratory, and with it, Watson's rats.[7]

While the physiological now trumped the behavioral at Meyer's Phipps clinic, Watson had switched the points: psychological research was steered away from the mysticism of Freudian interpretation and onto more rigorous and methodologically transparent tracks. Disenchanted with academia, Watson did not seek reappointment at another university, and instead accepted a job later that year with the J. Walter Thompson agency on Madison Avenue, where his program for the prediction and control of behavior found a more welcoming home: advertising. Watson remained a Mad Man until retirement.

RICHTER'S FIRST experience with rats came a year prior to Watson's ignominious departure, while he was still casting around for a subject for his doctoral thesis. He had returned to his rooms to discover that his supervisor had left him a gift: "I found a cage with twelve rats in my room. I was not sure whether Watson had sent them to me to give me some ideas about what I could do for a thesis, or simply to give me some company while I was making up my mind." Unsure what to do with his new rats, Richter contented himself watching them going about their business, "jumping around and climbing the walls of their cages." As he did so, a surprisingly simple question occurred to him: Why were they doing something rather than nothing? On what bases were the animal's apparently spontaneous decisions being made? Where Watson's behaviorist approach had been to train his rats, to condition their responses and control their actions, Richter took the opposite tack: "My interest focused entirely on what animals do on their own, not on what they can be taught to do." [8]

Studying the apparently random scurrying, patterns became discernible: regular intervals between periods of rest and activity. He timed the intervals, discovering rhythms that would later become the basis for his work on biological clocks. Richter adjusted variables—temperature, food, lighting—and noted how the rats adapted their behavior accordingly. He built what he called a "cafeteria" from which his rats could choose their own diet, and found that rats deficient in one or other nutrient would select foods that compensated for that nutrient, concluding that "the rat's appetite is a good guide to its nutritional needs."[9] In 1921, Richter submitted his doctoral thesis on "The Behavior of the Rat," and over the ensuing decades he would devise and conduct a series of experiments to investigate the biochemical and behavioral mechanisms by which rats were able to regulate their dietary requirements.

Rats proved to be scrupulous dieticians. In an experiment using alcohol, it was found that when offered a solution of five percent ethanol (about the same as beer) alongside their water, nearly all rats preferred the alcohol, and would drink this liquid exclusively with no apparent ill effects and without developing an addiction, unlike many humans. However, also unlike most humans, they treated the alcohol as food, and would lower their consumption of solids accordingly. It appeared the rats could count their own calories.

Upping the concentrations, and comparing the ethanol preference in rats with human subjects, Richter found that "a small but significant minority of both species, including human children, show a preference for solutions of up to 50% alcohol."[10]

At the time, the concept of biological self-regulation, the ability of an organism to restabilize itself in the face of external changes, was an issue of vigorous enquiry. In the mid-nineteenth century, the French physiologist Claude Bernard noticed that although the environment in which an organism lives may be highly variable, the *internal* environment of the organism—everything inside its skin or shell—must remain quite constant in order for most of the biochemical reactions that keep it alive to operate. That balance is extremely delicate. Bernard realized that the ongoing survival of an organism was only possible because the interior environment—*le milieu intérieur*—was essentially sealed off from the exterior environment. Under this description, we don't live directly in the world; we live within a sac of fluid, protected from the world as if inside a spacesuit. Once Bernard had drawn attention to the importance of the interior environment, others began to investigate the mechanisms by which this extraordinary process of biochemical self-regulation was achieved.

EXPLORING THESE new avenues of scientific research in America was Harvard physiologist Walter Bradford Cannon, who introduced the term *homeostasis*—meaning "staying in the same state"—to describe this constantly coordinated and minutely governed exchange of chemical information. When war broke out in Europe in 1914, Cannon was already an eminent figure, a member of the National Academy of Sciences, Chair of Physiology at Harvard Medical School, and President of the American Physiological Society. As it became clear that the conflagration in Europe would not in fact be over by Christmas, Cannon appears to have reorientated his research interests to anticipate the consequences and experimental opportunities of what was developing into a brutal and protracted campaign. Having made his name publishing groundbreaking work on the regulation of the digestive system, he now turned his attention to the broader issue of the physiological correlates of emotional states. Specifically, Cannon was investigating the role played by a secretion produced in the *adrenal medulla*, a small node of modified nerve cells perched on top of the kidneys. In medical terminology, *ad-renal* means "near the kidneys." That secretion was a hormone, *adrenaline*.

Adrenaline, or epinephrine, had first been isolated in 1901 by Japanese American Takamine Jōkichi, and Adrenalin became a proprietary trademark of his employer, Parke-Davis Pharmaceuticals, which marketed it for asthma relief. Takamine earned a considerable fortune in royalties, and subsequent copyright disputes over the use of the trademarked name led to US pharmacists adopting the generic term *epinephrine* instead.

At a time when nerve activity was thought to be wholly electrical, Cannon was pursuing the idea that hormones might act as a chemical neurotransmitter.[11] The notion that a gland nestled deep in the viscera could play a role in emotion and behavior seemed a throwback to the medieval belief in humors, while the implication that we might literally have "gut instincts" went against the psychological sovereignty of the brain. But during his earlier work on the digestive system, Cannon had already shown how peristaltic waves in the intestines of laboratory animals would cease abruptly when the animal was frightened. He knew that there was a spectrum of physiological responses to acute fear or stress: the liver released sugar into the blood, digestion halted as blood shifted from the abdomen to the heart and lungs and limbs, and cuts would clot more quickly. He also knew these effects were present long after the associated neural activity had ceased. He began to suspect that adrenaline was activating an unconscious emergency response to environmental threats.

"It is a matter of prime importance," Cannon wrote in 1915, "to determine whether the adrenal glands are in fact roused to special activity in times of stress." The problem was that it was difficult to find experimental subjects to observe during periods of intense emotional excitement. Cannon had to settle for ersatz proxies: during a Harvard-Yale football match in 1913, he waited in the locker rooms to acquire biological samples from the athletes fresh from "battle" (he discovered their urine had elevated sugar levels, consistent with the release of available glucose, but little else). World War I was to change all that. Two weeks after the US entered the war in April 1917, Cannon enlisted as a medic, and in May 1917, at forty-five years of age, with no military experience and only very limited clinical experience, he sailed for Europe.[12]

Posted to a field hospital in France, Cannon was presented with a chance to study firsthand and in human subjects the effect on the body of profound injury, blood loss, and shock. The benefits of having an on-site theorist as acute as Cannon proved significant: relaying the results of his investigations directly to the medical personnel treating the injured soldiers meant that the transfer of new theoretical understanding to applied clinical practice—a process that would normally take many months or years—was happening almost instantly.

Crucially, Cannon was able to decisively establish that the release of adrenaline into the blood altered the emotional state of the subject and triggered a series of physiological changes to prepare the body for action, leading to one of two responses: "The emotion of fear is associated with the instinct for flight, and the emotion of anger or rage with the instinct for fighting or attack"[13]—thus supplying English vernacular with "the fight-or-flight response."

Cannon's discoveries built up an increasingly coherent picture of how the endocrine system acted as the physiological mechanism by which homeostasis was achieved. Information acquired by the senses about the external environment was used to adjust the interior environment in order to maintain stability. Hormones secreted by the system of endocrine glands were the chemical messengers that passed instructions on to the various organs in order to make the necessary adjustments: increase blood pressure, transfer oxygenated blood to the limbs, stimulate the sweat glands to cool the skin, release sugar from the liver to support all of the above. It happened automatically, without any conscious oversight. Cannon called it "the wisdom of the body."

Much later, Cannon wrote that the resilience of biological systems was paradoxically ensured by their fragility:

> Organisms, composed of material which is characterized by the utmost inconstancy and unsteadiness, have somehow learned the methods of maintaining constancy and keeping steady in the presence of conditions which might reasonably be expected to prove profoundly disturbing.[14]

Homeostasis meant that an organism would make as many adjustments as possible to maintain an even keel, no matter how rough the seas. Even in war, the body would—for as long it was able—maintain a pretense of normality, as if to conceal from the soul its own distress.

A PARTICULAR set of questions, and a particular set of tools to answer them, was now laid out. It was clear that Pavlov's experiments had two distinct streams of influence: on the one side, they had inspired Watson to explore the psychological mechanisms involved in conditioning, which had led to behaviorism. Allied to Bernard's *milieu intérieur*, and Cannon's work on shock and adrenaline, the activity of hormones in establishing a steady state of biological equilibrium had produced the concept of homeostasis and the fight-or-flight response.

Richter, uniquely, was influenced by both Cannon and Watson, and his work with rats would come to unite those two streams—the physiological and the psychological merging now, becoming unitary. Following Meyer, he described himself as a "psychobiologist."

Richter's innovation was to absorb Cannon's concept of homeostasis and extend it to action. Where Bernard and Cannon had been investigating the physiological processes of self-regulation, Richter was looking at behavior as a response to environmental conditions. Alongside the internal physiological processes necessary to maintain body temperature or heart rate, how an animal acted could also be understood as an attempt to maintain homeostasis, to get itself back to normal.

JACK CALHOUN

TURTLE FARMS

1917–1934

BORN to a schoolteacher and homemaker in 1917 in Elkton, Tennessee, John Bumpass Calhoun, known to one and all as Jack, moves from one small town to another—Winchester, Brownsville—in his youth. His early childhood is spent under wide skies amid cotton fields and creeks and streams carved through sandstone caprock amid groves of shortleaf pine and oak. His mother remembers him as a solitary and contemplative child, curiously self-controlled, preternaturally patient. After a horseback fall lays him up with concussion, he develops the habit of stillness, spends his convalescence in what she calls "watchful waiting." He is always outdoors. Fishing, skating, and, as he would later remember it, "swimming in anything from ponds to earthquake lakes, to swollen, muddy rivers which somehow we managed not to drown in."

The American population was growing fast, and urbanizing in step: when Jack was born, that population was balanced almost exactly between rural and urban. By his death, nearly eighty percent lived in cities. The urban world was not contained within the half-distant cities but extended its tendrils even into rural Tennessee: their need to be linked up was impacting the countryside, too. In 1926, nine-year-old Jack watches helplessly as a small forest of ancient oaks on the edge of town was clear-cut to make way for the new federal highway from Nashville. When he looks back, it is the roads he remembers. They are the cities' protheses, reaching out across rural America, seeking sustenance for their destructive growth, metastasizing the urban model like a cancer.

He excels at school, but nature is his playground and his library. He channels his obsession with the natural world into hunting, cultivation, observation. At eleven, he is given his own small-bore shotgun. He becomes a driven collector: of birds' eggs and box turtles. He grows vegetables, which he sells door-to-door, using the proceeds to buy a sow, from which he breeds a litter of piglets. In his backyard, he builds what he calls a "turtle farm"—a miniature terrapin habitat, through which he diverts a stream to provide running water. He builds a terrarium in a glass tank, stocked with a snake he has trapped. He acquires a pair of field glasses to spot birds. Tennessee, the most biodiverse inland state, provides him with an inexhaustible supply of fauna to catalogue.

When the family moves again in 1930, it is into the city. They live in Nashville, amid the campus buildings of Vanderbilt. Jack's father introduces him to the Tennessee Ornithological Society. Under their guidance, Jack begins a more structured and rigorous regime of field observation, luring, trapping, birdbanding. He is taught how to prepare specimens for display, he helps out in the laboratory. Although the fieldwork is numerical and empirical, focused on populations and patterns, he is increasingly drawn to the behavioral idiosyncrasies of each species, each specimen: "Perhaps most important," he remembers, "I learned to know many birds as individuals, each with their own personality."

He's physically fit, having become an accomplished gymnast—a sport that suits his compact frame. When the Ornithological Society embarks on a project to band chimney swifts to map their winter migration, it's

Jack who climbs the chimneys to find their roosts. He is nicknamed "Muscle" by schoolmates who, when he graduates junior high, vote him "most likely to succeed."

Muscle Jack, Jack the tumbler, scaling chimneys in search of swifts. Jack with his backyard breeding sow, his terrapin colony. Jack birdbanding, swimming in swollen rivers and earthquake lakes. Eleven-year-old country boy with his own small-bore shotgun. Jack-all-alone on a ribbon of country road.

CHAPTER 3

BALTIMORE

CURT Richter hadn't set out to devise the perfect rat poison, but he was a restless experimentalist, and—unfortunately for his rats—the question eventually occurred to him: "If it is true that rats have such a remarkable ability to select beneficial foods and to avoid harmful ones, it may be asked how it is possible to poison them."[1] To this end, he began experimenting with compounds of arsenic, morphine, and mercury, and discovered that rats would avoid ingesting foods laced with even minute doses, far below the levels that could cause any harm. And then he heard about a curious property of a chemical called *phenyl thiourea*.

In 1931, research chemist Arthur L. Fox was working at the DuPont laboratories in Wilmington, Delaware, where phenyl thiourea (PTU) was being used as an accelerator in the vulcanization of rubber. While pouring a finely powdered preparation of phenyl thiourea into a bottle, some of the dust drifted out into the room. Fox reported what happened next:

> Another occupant of the laboratory, Dr. C. R. Noller, complained of the bitter taste of the dust, but the author, who was much closer, observed no taste and so stated. He even tasted some of the crystals and assured Dr. Noller they were tasteless, but Dr. Noller was equally certain it was the dust he tasted. He tried some of the crystals and found them extremely bitter. Naturally, a lively argument arose.[2]

As phenyl thiourea is fairly harmless to humans, Fox began to experiment: first on family and friends, later more systematically, to see who could taste the substance and who could not. He found only forty percent of people could detect the bitter flavor. Subsequent testing within families revealed that the trait was strongly heritable, so much so that before the advent of modern DNA testing, the ability to taste phenyl thiourea was even deployed to settle paternity disputes.[3] Fox had stumbled upon a hitherto unrecognized means of dividing the human population.

Richter became interested: Would rats also display a variable ability to taste the compound? He selected six specimens from his lab and offered each "a minute amount of phenyl thiourea powder—as much as can be put on the small end of a toothpick,"[4] and observed whether they tried to wipe it off their tongues and paws, as they would with bitter substances. When none of the rats reacted, he assumed they couldn't taste it. Arriving in his lab the next morning, Richter was astonished to find all six were dead. Postmortems revealed their lungs were swollen and filled with fluid. The compound was apparently tasteless to the rats, and yet lethally toxic even at very low doses.

In January 1942, Richter, now appointed head of the Rodent Control Project, set out to test his new poison in the field. Initial results were disappointing. Depositing small piles of PTU mixed with breadcrumbs in areas of high infestation, he found the wild rats were wary, unwilling to take the bait.

In the human gut, cells in the stomach lining are constantly sampling ingested foodstuffs for potential toxicity. If something sufficiently noxious is detected, a signal is relayed to the central nervous system and the vomiting reflex is triggered to expel the stomach contents. Rats lack that trip wire. Like horses and rabbits, rats are unable to vomit, and they are consequently hypervigilant about new foods. Hence, although they eat from the trash—precisely because they eat from the trash—they do not do so indiscriminately. And perhaps the phenyl thiourea wasn't entirely tasteless to the wild rats.

Richter set out to trap and collect rat specimens from the alleys and row houses around the hospital, and he established alongside his lab rats a parallel "wild rat" laboratory in the Carnegie Building of Johns Hopkins Medical School. Charged with collecting specimens was John T. Emlen, Jr., an ornithologist and naturalist from Pennsylvania. Emlen would not have chosen to work as an exterminator, but the war shuffled many into roles they hadn't anticipated. After registering as a conscientious objector, Emlen had been assigned under the new Civilian Public Service scheme to Richter's Rodent Control Project. The presence of such a competent naturalist was a boon for the project. Richter described Emlen as an "expert in population dynamics" whose particular skills proved vital for trapping and handling the specimens they caught, even if he probably spent more time studying the rats than trying to kill them.

As Emlen's haul accumulated in the Carnegie Building, the team quickly discovered the wild rats were a very different animal from their domesticated laboratory stock. Unlike the docile albino lab rats, the street rats didn't care to participate in experiments nibbling from toothpicks, and this presented a challenge for the scientists, who struggled to contend with their "aggressive fighting spirit" and "high intelligence." To minimize the bites they were all receiving, Richter constructed specialized leather funnels to handle the animals, using the wire ribs from an umbrella to fashion a collapsible restraint device. Even so, "we had to be on guard every second," he recalled, "which made it very exciting."

Where his previous work had investigated how rats adjusted their diets to maintain their health, Richter now sought to do the opposite: to discover a compound that would, as he put it, "trick the rats into eating substances that will destroy them."[5] From DuPont, he acquired samples of some two hundred different thiourea compounds and set to work testing these on the wild rats. Eventually, he found a variant that was highly lethal and almost tasteless even to the wild rats: *alpha-naphthyl thiourea*, or ANTU for short.

Synthesis of ANTU now began at pace—although throughout the espionage-haunted paranoia of the war years, the exact chemistry was to remain a closely guarded secret—and by the middle of 1942, they were ready to roll out the first phase of poisoning. To get around the wartime labor shortage, Boy Scouts were enlisted to distribute ANTU-laced cornmeal in an eight-block area around the university hospital. By October, they had expanded the operation to two hundred blocks.

The city block wasn't simply a convenient unit for administrative purposes, though. As they turned their attention away from the lab rats and began to study the behavior of wild rats in the city, it became apparent that the rats were territorial to individual city blocks. While Emlen was trapping the wild rats, he had noticed that each block had its own colonies, with almost no traffic between them. Those roads that had been laid out on the Poppleton map were acting like fences now, containing separate populations in discrete quarter-acre ranges. All the better for their extermination: when one block was eradicated, rats from the adjacent block would seldom cross the road to repopulate the territory.

As the Rodent Control Project was allied with the ongoing Baltimore Plan for civic improvement, Richter's exterminators were aided by the police force, who assigned a Sanitation Squad to accompany designated "Sanitarians" to patrol slum areas in search of unkempt properties and code violations. Two hundred thousand copies of a pamphlet called *You Can Help Fight The Rat* were distributed,[6] while a bombastic educational film, *Keep 'Em Out*, urged residents to "Fight rats with knowledge."[7] Straining to be plainspoken, and fashionably couched in martial register, the film deployed what the Johns Hopkins' scientists had learned about rat behavior to enlist Baltimoreans in the war against the rat: "It is always better to know the enemy before we begin to fight: by studying his habits and needs, we discover the weak points of his defense and then we can strike where it will do most good." After scolding homeowners for providing rats with "everything but the welcome mat," a three-step process for rodent eradication was laid out: "Block their entrance into buildings; Eliminate enclosed spaces; Catch and kill." Per the Baltimore Plan, the emphasis was placed on the individual: "*Rat control is up to you.*" As residents and landlords began (sometimes reluctantly) ratproofing their properties, Richter's Rodent Control team were gearing up to follow the Sanitarians, moving block by block, armed with industrial volumes of ANTU.

Abetted by the city's Sanitarians, the Rodent Control Project was by 1943 in full swing. In June, Baltimore mayor Theodore McKeldin gave Richter an office at City Hall and signed off on a no-strings-attached grant. "We are giving you $25,000," Mayor McKeldin told Richter, "and all we ask is that you kill $25,000 worth of rats."[8] A network of baiters spread ANTU in a coordinated effort over thirteen hundred and sixty city blocks. Residents were instructed to unlock gates and cellar doors to allow access for the Rodent Control Project, and fines were handed out for noncompliance. In some blocks, one hundred percent of rats were eliminated. In others, only ninety percent. Confident in the efficacy of his poison, Richter blamed the discrepancy on the "motley bunch of intelligent and not so intelligent characters" that were available to help out. That motley bunch—many of whom were volunteers—also targeted the city's factories and warehouses. In Lexington Market, over nine hundred rats were killed in a single night's poisoning. By 1946, ANTU had been used to treat over five and a half thousand city blocks, and—by Richter's estimate—well over a million rodents had been killed. Mayor McKeldin had gotten his $25,000 worth of rats.

With the war over, secrecy around the poison at the heart of the eradication drive was lifted. The Office for Scientific Research and Development, which had funded the Rodent Control Project, secured a commercial patent for ANTU in 1945, and the rodenticide was released for general sale in 1946. Advertisements declared "War-time secret revealed!" and heralded the arrival of "the miracle rat killer." Given that Richter had discovered ANTU's effects while investigating how the rat's appetite helps it to maintain a stable metabolism, there's a cruel irony to one such advert, which picks up on how the poison works by turning the operation of the animal's interior environment against itself: "They literally drown in their own body fluids!"[9]

YET AMID the fanfare, the ANTU poisoning campaign began to deliver steadily diminishing returns. Younger rats exposed to sublethal doses increasingly developed "bait shyness"—meaning that the poison became ineffective as the wild rats grew wise and could only be used every six months after new colonies eventually repopulated cleared blocks. Not only that, but the peculiar lethality for Norway rats didn't extend to other rodent species. Mice were unaffected, but dogs were, and orders to keep pets inside during baiting couldn't prevent accidental fatalities.[10]

Only two years after Richter's new miracle rat killer was released, a new poison emerged: *warfarin*, an anticoagulant that killed by slow internal bleeding. Tasteless and odourless, the rats were unable to detect its presence, while warfarin's action was sufficiently slow that they never learned to associate the poison with any particular bait. Consequently, rats would continue to return to warfarin-laced foodstuffs, slowly accumulating a fatal dosage. Death might take several weeks as they gradually bled out into their viscera. As warfarin proved its worth, ANTU fell out of favor as a rodenticide, and by the early 1970s it had been all but abandoned.[11]

Richter stepped down from his role as "a reluctant rat-catcher" in 1944, and temporary stewardship of the Rodent Control Project fell to John Emlen, whose skills as a naturalist and ecologist had been vital for livetrapping the wild rats. The decision to promote an ecologist was significant, as it had become clear that poisoning alone would not provide a permanent solution to the rat problem. While the ongoing project of urban rehabilitation slowly rolled out across more blocks, the majority of inner-city Baltimore still allowed the rats easy access to larders and warehouses, plenty of nesting places in attics and crawlspaces, and a buffet of trash in most every alley. In this habitat, the rats' high fertility rate meant they could replace their populations more quickly than Richter's team could poison them. So long as the city remained welcoming to rats, they would always return. A new approach was needed.

In 1945, with a grant from the International Health Division of the Rockefeller Foundation, the retitled "Rodent Ecology Project" was transferred from the Johns Hopkins Medical School to the School of Hygiene and Public Health, with population ecologist David E. Davis appointed as its permanent lead. As part of its philanthropic mission to improve global public health, the Rockefeller Foundation had previously employed Davis to locate host species of yellow fever in Brazil in the early 1940s, and he brought their trust and goodwill to the project at Johns Hopkins.[12] Under Davis, the emphasis would shift away from poisoning and toward rodent eradication through the application of environmental pressure: not now on the physiology of the rat, but on its ecology. Where Richter had hoped that a technological fix could be found, and focused his efforts on understanding the internal physiology of individual rats, the new approach operated at the level of populations: on interactions within and between colonies, and with their habitat. Their idea was that behavior was affected by environment, and if you could control the environment, you could control the behavior.

From Emlen's fieldwork tracking and trapping the Baltimore rats, Richter's group had noticed that they were reluctant to cross the roads and seldom strayed from the block in which they were born. That curious fact had made it easier to measure the success of the poisoning program. The Rodent Ecology Project took this observation as its starting point for a more focused investigation. Each city block was, as Davis put it, "effectively an island and its rats form a discrete population unit since immigration and emigration of rats is negligible or absent."[13] What invisible borders divided the populations? What mechanism kept them apart? To answer that, Davis was soon joined by a graduate student, John J. Christian, and a research assistant—a twenty-eight-year-old ecologist from Tennessee, recently matriculated from Northwestern with a Ph.D. on the behavior of the Norway rat: John B. Calhoun.

JACK CALHOUN

A STEEPLE FULL OF SWALLOWS

1935–1946

MARK Catesby, an ornithologist and illustrator from the village of Castle Hedingham in Essex, England, arrives in Charleston, South Carolina, in 1722. He has been commissioned by the Royal Society in London to catalogue birds of the New World. Among the species he records is the American Swallow, which he chooses to draw within a cutaway section of a chimney on whose curved interior wall the dun-brown bird grimly clings. A century later, John James Audubon calls it the Chimney Swallow, although it has by then already been reclassified as a swift.

The birds find themselves an unexpected beneficiary of European colonization. Before white settlers build chimneys, the swifts congregate in hollow trees to roost and breed during the summer, before heading south for the winter. There are many more chimneys than there are suitable hollow trees. And soon there are many more chimney swifts.

In the mid-1930s, the Tennessee Ornithological Society began a project of banding as many chimney swifts as possible, hopeful of one day solving the mystery of their winter residence. They know the birds flew south for the winter, but how far? And to where? In 1935, eighteen-year-old Jack Calhoun is apprenticed to the mission. His mentor is Amelia Laskey, who teaches him how to trap the swifts using a mesh cage, how to hold the birds and fix the bands. Laskey is an experienced birder but now in her early fifties, in no position to perform the perilous work of hauling a cage along rooftops.[1] Muscle Jack steps up.

In Tennessee that year, over twenty-seven thousand swifts are banded. The Nashville chapter records the highest total: 15,876. Ben Coffey, the project's coordinator in Memphis, writes with puzzled awe of "Mrs. Laskey's enterprising group of assistants being led by our aggressive and enthusiastic fellow member, John B. Calhoun, who must have been half fireman and half monkey to trap on the chimneys that he did."[2]

Scaling the sheer brick walls of factory chimneys to peer into the dim and soot-lined interior, Jack is astonished to find massed aggregations of many thousands of birds. The swarming was a consequence of the infrastructure: hollow trees had accommodated only a limited number of birds, but large chimneys were as cities, into which hundreds and even thousands might congregate. The population of swifts had never been constrained by their abundant diet of insects but by the availability of hollow trees of suitable girth. Within their cities, European settlers have built swift cities. It set small-town Jack to thinking: How can an animal accustomed to the company of perhaps a dozen others adapt to the presence of many thousands?

Much of what is known of the chimney swift has been discovered by Iowan ornithologist Althea Sherman. Barred by her sex from most professional bodies, obliged to quit her job as an art teacher to care for her ailing parents, she is a nominal amateur. To study birds, Sherman relies on the usual blinds and hides. But the chimney swift—nesting high in narrow shafts—presents a problem for observation. In 1915, she approaches two local carpenters with plans for an unusual project: a twenty-eight-foot tower with a hollow central chute, around which snaked a staircase. Small observation windows are cut facing into the chute. To the birds, it is a simulated chimney, as the chimney is a simulated trunk.

Duly, the swifts come, and for eighteen years she studies their annual arrival and nesting and departure, compiling the first detailed account of their life cycle in four hundred pages of notes. Sherman's swift tower is the first of its kind: an artificial habitat constructed purely for scientific observation. Is it still "nature observation," or is she studying the birds in captivity? The difference is moot. Sherman's swift tower is neither chimney nor tree. It simulates the environment in which the swift now lives. Like the Norway rat, the chimney swift has become a commensal animal whose life cycle is now tied to human habitations and changed utterly by it.

NASHVILLE, SUMMER of 1935: Jack is unexpectedly offered a ride to Cape Cod by a neighbor. He accepts a lift as far as Charlottesville to meet up with his friend, Jack Hayes. Hayes's father is a sociology professor at Vanderbilt but teaches summer school at the University of Virginia. Walking over the campus lawn, Calhoun and Hayes cross paths with renowned ornithologist Ivey Foreman Lewis, recently appointed dean of the university. Jack and Dean Lewis talk birds. The chance encounter proves significant: Lewis says the biology department is looking for a skilled collector; would Jack like a job? Calhoun agrees and is appointed on the spot. Lewis arranges to fund a scholarship.

When Jack begins college the following fall, he is both staff member and student. He has his own research cubicle, keys to the building, and, best of all, use of the departmental station wagon—"a privilege," he later recalls, "not shared even by the graduate students." The hybrid role accelerates his education: he attends staff parties and research seminars, and he is brought along for faculty field trips. In early summer, he is leased out to the Smithsonian in Washington, DC, where he works for ornithologist and deputy director Alexander Wetmore.

In the December 1938 edition of the *The Migrant*, the journal of the Tennessee Ornithological Society, twenty-one-year-old Jack Calhoun publishes his first scientific paper. He describes how, on the evening of August 30, swifts were seen circling above Father Ryan High School in Nashville. In flocks of up to a hundred at a time, they arrive low and fast, swooping over the silhouetted buildings. As dusk gathers, he watches as at least two and a half thousand swifts enter the chimney. The following day, he returns with ladders, ropes, and mesh boxes. On August 31, he logs a total of 4,467 birds.[3] That fall, when the birds fly south, he heads back north to Virginia for his final year. Shortly before graduation, in May 1939, one of the doctoral students tells him that Northwestern is looking for teaching assistants, suggests Jack apply. Northwestern wants a master's degree; Jack hasn't quite got a bachelor's yet. He applies anyway. In September, he is on his way to Evanston, Illinois.

This is the busiest place Jack has ever been to, but he spends little time in the city. Although the skyscrapers of downtown are just fifteen miles south of Evanston, he is drawn west, away from Lake Michigan, where "the whole area within a 100 mile radius of Chicago had become a living laboratory." Jack and his classmates explore the cycle of energy between upper canopy and leaf mold in Carle Woods, and the ant species *Formica ulkei*, "that builder of giant mounds housing huge cities." They compare the ecosystems of the glacial moraine surrounding the southern tip of Lake Michigan with the adjacent Indiana Dune country, long since lost under suburban sprawl.

At Northwestern, Jack is among what he calls the "core and acme of the Chicago school of ecology": Warder Allee, Alfred Emerson, brothers Orlando and Thomas Park, and Karl Schmidt. At Allee's house near the University of Chicago, the five professors meet with their graduate students for an out-of-hours Ecology Seminar every two weeks. These discussions will later form the basis of a landmark book, *Principles of Animal Ecology*. His time among the Chicago ecologists impresses upon him the interconnectedness of things, how nothing can be understood in isolation. Afterward, he will "no longer ever again view any event or process out of context of an expanding mesh of relationships affecting it."

Jack's supervisor, Orlando Park, works on how different species cope with sharing the same space. Alfred Emerson is a termite expert, a protégé of William Morton Wheeler and fellow proponent of the insect colony as a superorganism. Emerson investigates how termite nests instantiate and articulate their social organization, describes the structure of the termitary as "frozen behavior."[4] Orlando's brother, Thomas, is an experimentalist. His teacher, Raymond Pearl, got him hooked on population, and Thomas Park now builds miniature worlds. Sealing two species of flour beetle in a test tube with food and water, he watches as—each time the experiment is repeated—one species eventually outcompetes and drives the other to extinction. Schmidt is a herpetologist who has travelled the world in search of reptiles and brings to the group a global perspective on habitat diversity.

Warder Clyde Allee holds the intellectual center. A Quaker and pacifist, Allee is temperamentally opposed to the idea of nature red in tooth and claw. If competition is the central fact of life, war is inevitable. When he looks at animal populations, he sees cooperation and mutually beneficial behaviors. He observes how animals acting together confer shared benefits: through the choreography of pack hunting or defensive grouping, or by altering their own habitats with rabbit warrens or termite mounds or beaver dams. For Allee, the central mystery is aggregations: If survival is an individual task, why does sociality emerge at all? Allee's focus on group selection will later be widely repudiated, replaced by George C. Williams's totalizing account of what Richard Dawkins christened "the selfish gene." Just as under Thatcherite-style neoliberalism there is no such thing as society, only individuals; so under the neo-Darwinian description, groups are an epiphenomenon, genes do it all. But for Calhoun, the problem of social aggregations remains unanswered, gnaws at him.

Under Park and Allee's tutelage, Jack works toward a dissertation on twenty-four-hour cycles in the behavior of Norway rats. The work appeals to Curt Richter at Johns Hopkins, who will later be among the panel selecting researchers for the Rodent Ecology Project in Baltimore. During his doctoral study, Jack is the teaching assistant for an undergraduate class in field zoology. One of his students is Edith Delight Gressley. A minister's daughter originally from Pennsylvania, she's at Northwestern on a music scholarship, and will graduate a biology major.

They are married in Sandusky, Ohio, August 1942. Jack's ongoing research and teaching post grants him deferment from the war draft. The Calhouns move first to Emory, Georgia, then Columbus, Ohio, where Jack takes a teaching position at the Ohio State University. There he reads Charles Elton's *Voles, Mice, and Lemmings* and, following Elton, uses trapping data from the Hudson Bay Company records to model gene frequencies in fox populations.

But he dreams of modelling a mammalian population under controlled conditions: he thinks of Emerson's beetles battling for control of a test tube, of Althea Sherman's tower, which was neither chimney nor tree. Of clouds of swifts massing in the skies above Nashville.

In August 1944, the Fish and Wildlife Service in Washington publish a report concerning thirteen banded chimney swifts among a haul that have been killed by Indigenous hunters during December 1943, at Yanayacu in Peru, where the confluence of the Ucayali and Marañón rivers forms the head of the Amazon. The bird bands are presented by the hunters to an American from the National Library and find their way to the American Embassy in Lima. For the first time, the site of the wintering is known. The dead birds were banded in Georgia, Alabama, Illinois. One came from as far as Ontario. Eight were from the Tennessee survey. Two were logged to John B. Calhoun, one of which was dated "8-31-38"—the day of the mass capture at Father Ryan High, five years earlier and some three thousand miles north.[5]

CHAPTER 4

RODENT ECOLOGY PROJECT

WITH ecologists at the helm, the mission of tackling Baltimore's rat problem now took a markedly different turn. As befits the name, the Rodent Control Project had sought to control the rodent problem in the city, and control the rats in the lab. To this end, Richter's rat room at Johns Hopkins had been pristine: each wall blocked into neat rows and columns of identical cages arrayed floor to ceiling, like the atrium of a modern office building, or a prison hall. Gleaming and clinical, it produced systematic and orderly results. Here, the rat could be studied in a controlled and standardized manner, experimental findings measured and tabulated with precision.

When Davis arrived at Johns Hopkins in 1945, he was happy for Richter to keep his rat rooms. Davis wasn't especially interested in individual cages because he wasn't especially interested in individual rats. His area of expertise was ecosystems: the interaction between populations and their environment. The important thing for Davis was the interaction. He thought it was impossible to divide ecosystems into logically distinct components—the populations on the one hand, their environment on the other—holding that the "division into forces and factors is artificial because each group interacts with the others." He was particularly dismayed by the direction taken by the burgeoning field of mathematical population dynamics, which sought to chart and predict the growth of a population using mathematical models based on cell cultures grown on a petri dish or fruit flies in a bottle. These were living populations, but highly abstracted: the organisms were merely demographic ciphers, chosen precisely because their growth could be easily modelled.

Anticipating what would become a familiar critique of computer simulations today, Davis felt that such models were self-satisfying, producing circular results that were only as good as the assumptions they started with. And although he stopped short of calling the modellers themselves self-satisfied, he clearly felt they designed experiments that would produce the sort of results their models predicted. "In the opinion of some investigators," Davis wrote, in an exemplary display of authorial distancing, "this approach has only limited usefulness because it merely confirms statements about populations in very simple situations and does not reveal new truths apart from those revealed by the mathematicians." He preferred the inductive method of the field ecologist: amassing reams of empirical data on the movements, mortality, and reproductive rates of a population, meticulously cross-referenced against the environmental conditions. He happily conceded this data was "often uncontrolled and frequently undigested" but said it had the virtue of capturing "what actually occurs (within limits of error) under the complex conditions of nature, and thus may reveal new truths."[1]

Throughout Richter's poisoning campaign, Emlen had been conducting his own studies on the movements of the rats within and between blocks. Handing over stewardship of the operation, he now presented Davis with the preliminary results of his fieldwork, showing how rats seemed to stay within an area as if tethered to an elastic leash. Davis was fascinated, and the two men set about collecting as much information as they could about precisely how far the rats travelled, and which animals went where. Richter had brought the wild rat into his laboratory, now Davis would turn that investigative model inside out: he would make his laboratory the city itself.

Rats had lived in American cities as long as people had: either both belonged there, or neither did. Davis was one of the first ecologists to treat a city as a natural habitat, and to apply the fieldwork techniques of ecology to commensal animals in the urban environment. The trapping and surveying techniques he and Emlen developed in Baltimore would subsequently debunk the popular factoid that held that in cities there was one rat per person—a ratio that would yield what Davis called an "absurd estimate" of eight million rats in New York City.[2] By his calculations, the figure was closer to a quarter of a million.

Joining them was Jack Calhoun, and John J. Christian. They were almost exact contemporaries, Christian just four months older. As a nine-year-old, Christian had been laid up for months with a critical illness and—much like Calhoun—had spent the still days of his convalescence watching birds in the maple trees outside his house. He developed a love of nature, went on to study biology at Princeton, and was training to be a doctor at Columbia Medical School when Pearl Harbor was attacked in 1941. He switched to aeronautical engineering, enlisted in the navy, and served on a torpedo patrol boat in the Philippines, where he spent his spare time studying the habits and drawing illustrations of the fishes and birds of the South Pacific. After the war, he decided he preferred working with animals and took a post with the Pennsylvania Game Commission surveying populations of small mammals, which set him on the road to a career in vertebrate ecology and led to his being selected for Davis's team at Johns Hopkins. When Christian asked his new boss if he was more interested in birds or mammals, Davis promptly replied: "Neither. Populations." Davis set the tone for the Rodent Ecology Project. A highly capable administrator, he encouraged the team to apply an organized and disciplined approach to their studies. The city might be chaotic; their work would not be.

One of the first things they did was to release wild rats back onto the streets. Trapped rats would be "marked"—a process that might have been accomplished with a dab of paint, but usually involved clipping off a few toes—and then released back near the same block. A daily process of trap-and-release would then be implemented, noting if any of the previously marked rats were recaptured and, if so, how far away from their point of release. This was intensive work: during one six-day period, one hundred and twenty-eight rats were marked and released, and sixty-two of these were recaptured one hundred and eighty-six times. Sure enough, the rats remained strictly within their own neighborhoods: none were found farther than one hundred and twenty feet from their point of capture, and only three more than eighty feet away. Over four hundred rats in adjacent blocks were trapped to see if any of the marked group had strayed. None were ever found to have done so.

The fieldwork of the Rodent Ecology Project was extraordinarily thorough: block by block they trapped and released, keeping count of population sizes, how many dead rats they find, how many litters of new pups are born. When it snowed, they used the opportunity to track individual paw-print trails, carefully marking these onto three hundred and fifty hand-drawn maps. During drier months, they deposited food containing a dye that would stain feces blue, and then searched for traces of the blue rat droppings to chart the routes of the most popular rat runs.

Davis and Emlen were building up a picture of the so-called *home range* of the Norway rats. The home range is the area around the homesite or nest within which any representative individual will usually roam in search of food. So while squirrels and mountain lions might be found throughout the Pacific Northwest, each mountain lion might have a home range of one hundred and fifty square miles, while the squirrel a home range of only a few acres. Today, home range is a key ecological concept, but when Davis and Emlen began their survey of Baltimore rat movements, it had only recently been defined. In an essay from 1943, zoologist William Henry Burt had distinguished home range from *territory*: territory was just that subset of the home range that the individual would defend against intruders. Your house would be your territory, your neighborhood your home range. Burt stressed that while only some species had territories they fought to defend, every mammal had a home range, the size of which would be characteristic of that species.[3]

By checking the locations of marked and trapped rats, Davis and Emlen were able to work out that the home range of the Norway rats was between one hundred and one hundred and fifty square feet. To check if this was a peculiarly urban phenomenon, they extended their fieldwork to a local farm where, despite being surrounded by open fields, individual rats confined themselves to similarly limited territory—all the rats cleaving close to the various outbuildings. Not only that, but rats from the stables stayed away from the barn, and rats from the barn stayed away from the stables.

They discovered that the rats lived within small colonies of ten to fifteen animals and, in the city, there were ten to twelve colonies per block. Within each block the total population remained fairly constant, usually around one hundred and fifty rats. When the numbers were reduced by poisoning or trapping, these block populations would steadily return to their equilibrium point. This went some way to explaining why they never crossed the roads: they didn't need to. Their colonial domain was already smaller than any individual block.

The stability of population numbers within blocks was quite robust: it didn't seem to matter what happened to the rats, they bounced back to approximately one hundred and fifty per block. The project team already knew that poisoning had an only temporary effect, but what about predation, or disease? Were the local cats and dogs aiding the eradication program? Not so much. Those who bought cats to keep out the rats had been sold a dud. "Cats," Davis noted, "prefer to eat garbage." Even keen feline hunters could only dispatch about thirty rats per year, a negligible dent in the rodent population, and, because any rats the cats were able to catch were usually already sick or old, these killings had little impact on reproductive rates. If anything, they freed up more resources for the healthy, who then had more babies. As Davis put it, predation "merely reduces the population to the level of maximum reproduction and causes the maximum rate of conversion of garbage into rat flesh."[4] A study from Germany found the presence of cats actually increased the rat population: "The conditions that bred cats also bred rats." Dogs were even less useful: food left out for dogs actively encouraged the presence of rats, and the largest rat they ever caught—weighing almost twenty-five ounces—was found under the kennel of a well-fed dog.

Wild rats were also surprisingly disease resistant: although many trapped specimens harboured roundworms and *Leptospira* bacteria, there was no corresponding sickness. The parasites seemed to have struck a deal with their hosts to guarantee the ongoing survival of each. *Salmonella* infections, however, had been observed to bring down populations, so the team briefly tried germ warfare: infecting trapped rats with Salmonella and then releasing them back into their colonies. Again, the results were only temporary, and the rapid reproductive rate—each female producing as many as forty pups per year—meant that populations bounced back just as swiftly as they had following a poisoning drive.

Jack Calhoun raised a counterfactual: Would it be possible to *increase* the population of a block? Given that rats didn't voluntarily move between blocks, they would need to be relocated involuntarily, but there was plenty of unused space and abundant food. So they captured rats from around the neighborhood and introduced them into one of the study blocks. "Perhaps," Calhoun thought, "if I dumped in a large number of strangers they would establish a second layer of rat society."[5] But when they came back to count, the new rats had all vanished. There was no second layer of rat society. Christian said they had "become mortality statistics."[6] The block population certainly didn't increase, but nor did it return to its previous levels. Instead, it decreased, and significantly: Christian was later able to calculate that adding twenty percent more rats to a stable population would cause the population to contract by sixty percent. This was unexpected: it looked like the most effective rat killer was more rats.

Both Christian and Calhoun suggested that the presence of new rats led to "social strife" as graded dominance hierarchies were disrupted by fighting and competition. "All hell broke loose," Calhoun recalled. "Strife among strangers, strife between strangers and residents, then strife among residents."[7] Davis wrote of the "psychological turmoil" caused by the introduction of alien rats into the territory of a stable and "happy" rodent colony where "everybody knew who was married to whom, and whose children were whose, and so on."[8] The forced immigration policy seemingly caused problems from which block communities could not easily recover. The idea that mere social disruption might cause a more durable drop in population numbers than poisoning, predation, or disease was a remarkable one.

No one was really sure what was happening. It was Christian who cracked the problem. He had a deeper knowledge of physiology than the other members of the team, and postmortem examination of rats from the forced-immigration blocks revealed symptoms that looked familiar to him: they had swollen adrenal glands, their lymphatic system was shrunken, and they had ulcerated stomachs. While a medical student, he had been a lab assistant in Columbia's anatomy department, working among pioneers in the new field of endocrinology on the pharmacology of *corticosteroids*. In his bacteriology lectures, he had learned that mice immunized against Salmonella forfeited that immunity if kept in crowded conditions. Seemingly, high population density compromised their immune systems. The year before moving to Johns Hopkins, he had read a paper by Hans Selye on the effects of *adrenocorticosteroids*, a hormone produced by the adrenal glands under conditions of environmental duress.[9] When Christian remembered the Selye paper, everything snapped into place.

JÁNOS HUGO Bruno "Hans" Selye was born in Vienna in 1907 during the heyday of the Austro-Hungarian empire and emigrated in 1931 to the US, where he first worked at Johns Hopkins before moving north of the border to take a permanent position at McGill University in Montreal. During his medical training, Selye had become fascinated by something so obvious that no one had really thought to investigate it: he noticed that all the patients in the hospital felt sick.

Whether they had cancer or tuberculosis or gastroenteritis or a broken leg, all had the same background symptoms: they were listless, sedentary, apathetic. They were off their food, they didn't feel like doing anything. Along with their specific medical problem, they simply felt unwell—but what struck Selye was that everyone felt unwell in the same sort of way. Selye called it "the syndrome of just being sick." He framed his question more sharply: At the physiological level, what is happening when a patient is "feeling sick"? That this seems so simple is precisely why it had for so long gone unnoticed, not unlike Richter's question of why his rats were doing something rather than nothing.

Selye suspected that what patients were experiencing was the effects of the body trying to adjust to its new condition—the broken leg, or the tumor, or the tuberculosis infection. To investigate what underlay this nonspecific and seemingly universal reaction to adverse conditions, he conducted a series of experiments designed to make rats feel unwell. Some were injected with chemicals such as formaldehyde and morphine, or were exposed to cold to induce torpor, or were secured within motorized exercise wheels to induce muscular exhaustion. Some had their spinal cords severed. But regardless of the particularity of the ailment, over time, all displayed the same suite of physiological responses: enlarged adrenal glands, an atrophied lymphatic system, and peptic ulcers in the stomach and duodenum. It was essentially this triad of complaints, Selye contended, that accounted for the feeling of generalized distress.

Announcing his findings in a short note in the journal *Nature* on July 4, 1936, Selye described how "a typical syndrome appears, the symptoms of which are independent of the nature of the damaging agent or the pharmacological type of the drug employed, and represent rather a response to the damage as such."[10] He outlined three distinct phases following the infliction of harm: the first was a "general alarm reaction" in response to suddenly being confronted with a critical situation, marked by rapid changes in glandular activity and metabolism. The second phase began about two days later as the body recalibrated its interior environment to adapt to its new predicament. During this "general adaptation syndrome" phase, the organism became habituated to its discomfort, and the internal organs and glandular secretions returned almost to their normal operations. An organism could sustain resistance for a long time in the "general adaptation" phase, but if whatever was causing the reaction continued, the body was unable to maintain homeostasis and would eventually lapse into a terminal phase that led to exhaustion and death.

Selye would later call this physiological response *stress*, and whatever caused stress was a *stressor*. The concept of stress was already floating around (Cannon, for example, used the term in 1915 when enquiring whether the "adrenal glands are in fact roused to special activity in times of stress"[11]), but in demonstrating the body's passage through three distinct phases of alarm, resistance/adaptation, and exhaustion, Selye medicalized the word, describing precisely the somatic consequences of extended periods of discomfort.

Selye would go on to broaden and develop the concept of a stressor to include any environmental stimulus that produced this pattern of responses from the endocrine system, and to apply that to human populations. His popular 1956 book, *The Stress of Life*, brought Selye's ideas to a wider audience, and "stress" quickly entered common parlance. He would subsequently distinguish between stress that was damaging—*distress*—and stress that was potentially beneficial, which he called *eustress*. Eustress was stress that geared the organism for action and that could present physical symptoms such as nausea or diminished libido or anxiety as a prompt toward making lifestyle changes. Stress was, in a sense, an existential equivalent of pain: the mute wisdom of the body sending a somatic signal that something was wrong.

The stress response was a slow-motion version of an adrenaline rush. Where the fight-or-flight release of adrenaline that Walter Cannon had described in World War I soldiers was a very short-lived emergency response that largely exhausted itself in victory, escape, or death, Selye was interested in the effects of long-term, chronic stress. Rather than an adrenaline rush, this was more of an adrenaline trickle. But that slow drip gradually eroded the body's capacity to cope. Environmental factors that might be merely inconvenient in the short term—such as persistent loud noise, financial worries, job insecurity, marital discord—over time exacted a heavy price. And eventually, shading into the third phase, stress led to more serious conditions such as heart disease and high blood pressure, and ultimately contributed to persistent poor health and a truncated life span.

Just as Watson had operationalized behavior, converting it into an empirical output, a measurable variable, so Selye had translated the vague sense of disquiet and malaise into something that could be recorded with laboratory tests. In humans, this was trickier than it had been with the rats, as dissection to analyze the endocrine and lymphatic system was obviously not possible. But levels of the steroid hormone *cortisol*, a product of the adrenal glands, were higher in stressed patients; levels of a neurohormone called *catecholamine* were also elevated. High levels of catecholamines cause a pounding heart, sweating, anxiety, and raised blood pressure. Stress gave physiological substance to the psychological experience.

Christian would now posit that something similar was happening with the city rats in the forced-immigration blocks. The presence of alien rats among socially stable block colonies resulted in unwanted social interaction, which led to agitation. The agitation caused by outsiders wasn't immediately or directly fatal, as Richter's poison had been, but it might be sufficient to disrupt normal behavior, including reproductive behavior. So perhaps one consequence of a stressful environment was a lower birth rate. It could be understood as a ranking of priorities: reproduction was obviously important, but—unlike eating or drinking—sex could wait. This would also explain why the effect endured even after the alien rats had become mortality statistics: rats that don't breed don't bounce back.

Where Cannon had initially described homeostasis as a physiological issue, and Richter had extended that regulatory activity to include behavior of individuals, the Rodent Ecology Project now scaled that up another step, proposing a form of homeostasis operating at the level of the population. As Davis realized, if rats in a stressful environment suspended breeding activity, there was scope to interpret this as adaptive. Richter had shown that rats counted calories to regulate their nutrition. It now seemed they were also able to practice population control.

All this talk of rats struggling to reproduce successfully when "stressed" by the presence of outsiders might sound like a glib overlay of human societal norms onto rats. But it is important to note that the concept of stress had not been transferred from humans onto rats: the conceptual priority of stress lay in Selye's rat experiments. The biomedical understanding of "stress" was a finding derived from the study of rats and only latterly transferred to humans. Stress was as much a rat problem as it was a human one.

For his part in the project, Jack Calhoun had been focused on how the distribution of rats within each block might play a role in their social interactions. A typical block of row houses might have anywhere from twenty to seventy dwellings, arranged in two terraces back-to-back, serviced by a central alley. Rat colonies within the block would live as close as possible to a food source but as far away from one another as possible. Between them, the invisible envelope of their colonial home range acted like a magnetic field, holding members inside the territory and repelling outsiders. Jack hypothesized that the spatial arrangement was a factor mediating social competition and, per Christian's stress theory, regulating the size of the block populations. When Davis later summarized the work from the Baltimore operation, it's this issue of social organization with which he concludes his findings:

> The habitat is fundamental in the determination of population size, but the density-dependent factors (predation and competition) are the means of holding the population within the limits imposed by the habitat. For rats, competition as manifested by social organization is the more frequent mechanism for limiting the population and thus deserves increased attention.[12]

THEY KNEW the methods of environmental modification they were using were working: clearing trash, blocking entrances, and repairing fences reduced the habitable space and restricted access to foodstuffs, which increased stress, which in turn appeared to inhibit reproduction. Their efforts were so successful that Davis would later estimate that during the years his team was active, the rat population of Baltimore decreased tenfold: from four hundred thousand in 1945 to just forty thousand by 1948.

But they didn't really know any detail. While Davis could point to disruption to "social organization" as the probable mechanism leading to lowered fertility rates, and Christian had noted the enlarged adrenal glands as the physiological consequence, the part in between was unknown. They didn't know what disruption to rat society meant because, beyond the organization into several colonies of approximately twelve per colony, they didn't really know how rat society worked.

When the project had migrated from Rodent Control to Rodent Ecology in 1944, the focus shifted from eliminating rats to understanding how they lived. And although the team had learned so much about the Norway rats, they had now reached an observational horizon. Tracking footprints, counting block populations, and tracing the spoor of blue feces offered only a letterbox-view of their behavior; trap-and-release gave only occasional snapshots of activity. Richter had brought wild rats into his pristine laboratory but learned little of how rats behaved in the wild. Davis and Emlen had made the messiness of the city their laboratory, but what they could learn was stymied by what they could observe. A new sort of laboratory was needed if they were to learn more.

Jack Calhoun, who remembered Althea Sherman's false-chimney tower, who as a boy had built simulated environments for his terrapins, had a suggestion: they should make their own city block.

CHAPTER 5

TOWSON

ONE cold dry December day in 1946, on a dead-end track off the York Road just outside Towson, John O'Donovan was walking his two setters in the grounds of his estate. O'Donovan lived in a large old house amid almost five hundred acres of land, on which he owned several properties. Some months back, he had rented one of these to a young couple, Jack and Edith Calhoun, newly relocated from the Midwest for Jack to take up a position at Johns Hopkins. The Calhouns balked at the prospect of living in the city itself: "The row-house pattern of habitation characterizing most of Baltimore presented a face which Edith and I, with our small-town early backgrounds, would just as soon avoid." Towson sat just eight miles north of downtown Baltimore but felt a world away from the city. So the place here on the edge of O'Donovan's estate, some way from the town and backing onto a forested hillside, suited them just fine.

Noticing his new tenant out in the yard as he passed by, O'Donovan stopped to talk. He asked Calhoun about his work. Jack told him he was part of team studying the behavior of rats in Baltimore, but that they were now approaching the limits of what they could discover in the field. Jack said they needed a dedicated space where he might "set up a pen to house a few rats." Gesturing toward a clearing on the hill behind the house, Jack asked O'Donovan if he'd be willing to allow him to build an enclosure up there. "Sure," said O'Donovan, "go ahead." And with that, he turned to walk back home through the woods.

Two months later, O'Donovan next passed by with his dogs. He was shocked to discover a huge enclosure, a hundred feet along each side and containing what seemed be a set of concentric fences, with a high observation tower at one side. It looked like a prison. Calhoun said it was "an artificial city block."

The Towson enclosure covered a quarter of an acre, with four-foot-high wire walls extending a further two feet underground. To deter escape by climbing over or burrowing under this, a wooden shelf overhung the perimeter around the lip of the wire, and sheeting sealed off the subterranean section. Just inside the fence, arrayed nine per corner, were thirty-six partially submerged and numbered harborage boxes. On the southwest side, the twenty-foot-high observation tower loomed over the pen. To complete the penitentiary aesthetic, a three-wire electric cattle fence surrounded the entire setup (although this was intended only to keep predators out).

Everything about the Towson enclosure was designed as a full-scale replica of a Baltimore row-house block. Although the footprint of a real city block was considerably larger, most of that area was inaccessible to the rats because it was already occupied by building structures; a quarter acre approximated the space actually available. He recalled how Davis and Emlen had tracked "rat-runs" through the blocks and replicated similar structural obstacles. As rats in the alleys had to share the same pathways, so Calhoun had cleared "alleys" through the brush and shrubs to simulate the pathways and passages available behind the row houses and lined these with fencing to simulate the walls' dividing yards. And because all rats within a block would be obliged to feed from a limited number of trash cans, the food hopper in the Towson enclosure was situated in the middle, behind a quadrangle of fencing rotated forty-five degrees with respect to the outer perimeter, through which were cut just four access points. As in the city, the rats would need to leave the safety of their colonial turf and risk contact with rival groups to obtain food. And, as in the city, those access points would be bottlenecks where social contact would be more likely to occur. When O'Donovan first saw it, the pen wasn't quite finished. It didn't yet have any rats in it. But now, and with O'Donovan's reluctant blessing, "the settling of the 'new continent' began."

To recruit those first colonizers, Calhoun took the ferry east from Annapolis to Kent Island, following the route of what is now I-50 over the Bay Bridge. Early maps of the Chesapeake show a spit of land called Parson's Point extending out into Prospect Bay, just south of the Kent Narrows. At some point in the nineteenth century, coastal erosion severed the isthmus, and Parson's Point became Parson's Island. When Calhoun came out here in 1947, the rats of Parson's Island had been isolated from the mainland for perhaps a hundred years. He knew there was a native rat population here because—just the previous year—Davis and Emlen had been hired by Baltimore's McCormick spice company to poison the island's rats, bringing down the numbers from almost seven hundred to a little over two hundred. McCormick had acquired Parson's Island in 1944, hoping to use the land to grow spices when war disrupted imports. That experiment was a failure, and they left the land fallow, built a clubhouse, and kept Parson's Island as a wilderness retreat for corporate events. Quite unintentionally, McCormick had instead cultivated a game reserve for wild rats.

The limited opportunities for outbreeding, coupled with the reduction from poisoning—Davis and Emlen's intervention in 1946, and an earlier poisoning drive in 1923—meant that the population would have very low levels of genetic diversity. This was important: Calhoun needed wild rats that were as homozygous as possible so that any subsequent variability in growth or behavior observed in the new pen could be attributed to environmental factors, not heredity. In February 1947, he trapped five pairs and took them back to Towson.

Population studies had always been small affairs—using fruit flies in a jar, for example—or else they were field studies, and consequently limited in observational capacity. No one had ever attempted a controlled mammalian population study on this scale. Calhoun would later describe how he had envisaged nothing less than "a closed universe where a group of rats could set out unmolested to establish their own empire on such an artificial island."[1] After releasing the five seed pairs, he retreated to his tower and waited, intent on interfering as little as possible with his new population, so that "such changes as might occur in the colony should be brought about by the activity of the rats themselves."[2]

When birdbanding as a teenager, Calhoun had spoken of how he came to know each specimen as an individual. He would now apply that approach to the Towson enclosure with the diligence of a census taker, deploying all the bureaucratic machinery of civic governance: Every rat is individuated, every burrow mapped, every parturition recorded. Litters are counted, infant deaths subtracted. All deaths are registered, and, where possible, the cause. For twenty-seven months, Calhoun would stand in the tower with binoculars and notepad, crawl among the alleys he had cut and fenced in. Watchful waiting. Reams of tabulated data accumulate. Here was Janet Kemp's *Housing Conditions in Baltimore*, but directed at the rat city he had built.

After a little over two years of daily observations, and another six years analyzing the collected data and writing up the results, a first draft of his findings would be more or less complete by late 1954, although publication stalled for another nine years. When it finally emerged in 1963, Calhoun's monograph on *The Ecology and Sociology of the Norway Rat* was, and remains, the most comprehensive and complete account of rodent behavior ever produced. Calhoun knew more about the behavior of the Norway rat than anyone else alive.

The detail is quite staggering. Where Davis and Emlen were able to approximate the home range of rats in the city blocks, Calhoun could now chart the precise movements of individual animals from birth to death. The rats were each assigned a number, and their performance across their lives was studied minutely. Any fights and their outcomes were recorded, indexed against the age, weight, and social rank of the combatants. At death, they were dissected. In females, uterine scars were counted to measure embryonic fertility against recorded births.

As in the city blocks, the rats broke off into separate colonies distributed throughout the enclosure, where they entrenched themselves into burrow systems. How space within the enclosure was used became a matter of increasing interest. Discovering that the rats use two distinct burrow plans—a circular design with chambers arrayed in a ring around a central hub, and a linear design with chambers budding off a straight corridor—Calhoun noticed that which of these plans a rat occupies had a profound effect on its social life. In the circular burrow, he would write, "each inhabitant will be brought into frequent and nearly equivalent contact with every other inhabitant." But in the linear or elongate design, rats living at one end would seldom venture to the other, and hence "the farther apart in the burrow two individuals live the fewer will be their contacts."

One way to read this is that the linear design affords its inhabitants more privacy. Another is to say that the elongate burrow inhibits sociability, fostering the very isolation it affords. Because the elongate burrow compels a rat to walk the same route every day, it creates habit and routine, whereas the circular burrow creates choice—and choice creates complexity. Like little rodent existentialists, the rats who live in the circular burrows experience the "conceptualization of possible modes of action," and this in turn has "reverberations on the ability of the individual to develop complex social interrelationships." The design of their living quarters determines how they behave and what sort of social individuals they are able to become. A higher-order query arises: if circular burrows are beneficial, why are there different burrow plans in the first place? Because Calhoun has monitored every individual, he is able to track which rats live in which design, and over time, a pattern emerges.

DURING THE final year of the experiment, the Towson rats had settled into a fairly stable organization of eleven separate colonies, which Calhoun labelled *a* through *k*. These differed widely in their constitutions and the well-being of their inhabitants. Stronger colonies had more females than males, less immigration from other colonies, and higher reproductive success, and their members displayed fewer wound scars. In the group with the healthiest rats, colony *a*, a single dominant male—coded as "49"—bred exclusively with thirteen females. Male 49 was unquestionably the most dominant rat in the enclosure. He displayed the largest home range of any specimen, roaming the entire pen, and was observed in hundreds of fights, all of which he won. Overseen and protected by Male 49, the females of colony *a* had a high number of surviving young, robust bodyweight, and longer lives.

Colony *a* had a complex network of social integration and almost no migration from other colonies. The next three most successful colonies followed a similar pattern, with a small number of dominant males breeding successfully with a larger number of females. These were the circular burrows, and they were spaced out from one another with easy access to the central food hoppers.

Farther away from the food supply, the remaining colonies descended stepwise in both health and social coherence. The architecture of the burrows closely mapped their social structure. The elongate, linear burrows were really an expression of social disorganization, contributing as both cause and effect. When space was limited—when a fence, for example, blocked their expansion—an elongate burrow was more likely to form. But that structure, by inhibiting the opportunities for normal socialization, resulted in weaker group cohesion and lower rates of reproduction. There was also more population flux, as the less-dominant males were driven out of the high-functioning circular burrows, widening the differential sex ratios in the linear burrows yet further. Rats displaced from the colony they were born in, usually young males, would attempt to join an adjacent group, but the presence of too many newcomers caused disruption, fighting, and ultimately colonial disintegration.

Expelled from their native burrows, low-ranking males wandered around the pen in search of a dwelling but were aggressively turned away from any colony with more dominant males. These rejected males formed the very tail end of the dominance gradient. They eventually collected in an "all-male aggregation" of "social outcasts." These were the doldrums, colonies *j* and *k*. Occupying a deserted tunnel initially for "overnight residence," their burrow became a sort of halfway home, the rats within listless and beaten, their bodies laced with the scars of lost fights. A loose band of exiles, they lacked social cohesion and, with no common background or family ties, failed to construct linking tunnels between their nest sites. Unwilling or unable to structure or maintain their habitation, things fell apart. Very swiftly, Calhoun would observe, "the burrow began to decay and the tunnels and nest cavities became a stinking mire."

Despite being genetically almost indistinguishable, and having the same access to food, rats in the elongate burrows grew up to be physically smaller than those in the dominant colonies, more vulnerable to disease and infection, and they died younger. Isolated from females, they never reproduced. What these refugees gained from their ragged community is unclear, and Calhoun doesn't speculate on what drew them together: "To what extent rats of disturbed past history are attracted to each other is unknown." What was known was that social rank and living space were linked, and in concert affected both behavior and physical health. Low social status stunted development, shortened lives. What emerged was a systemic grouping of discrete colonies, arranged in order of fitness. Calhoun called it a "class structure."

The most striking and enduring finding that emerged from the Towson experiment was the appearance of a natural cap on population size. After the class-structured colonies had become established, the total population of the pen levelled off at around one hundred and fifty rats, and never exceeded two hundred. That figure was in line with the natural populations found in the Baltimore row-house blocks on which the enclosure was modelled, but fell drastically below what might have been expected if the rats of Towson had bred according to geometric progression.

Extrapolating from reproductive rates, with eight pups per litter and one litter every two months, during the two years of the study there should be as many as fifty thousand rats within the fences. In reality, such a figure would never be achieved outside of a crudely idealized mathematical model, but the final tally was still surprisingly low. The Towson enclosure covered about ten thousand square feet. If a laboratory was constructed to comparable dimensions, and each rat isolated within a standard two-foot square cage, it would be possible to accommodate at least five thousand healthy rats in the same area. What, Calhoun asked, could account for this twenty-five-fold decrease in the potential population size?

With the usual suspects of resource scarcity, predation, and disease eliminated, the only pressure now was internal competition—social strife between residents. When Calhoun had been working alongside Davis and Emlen in Baltimore's row houses, he and Christian had suspected that unwanted social interaction was acting as a brake on the rats' reproductive rate. The Towson experiment confirmed this and afforded Calhoun the opportunity to observe directly the mechanics of spontaneous population control.

Successful breeding that saw pups raised to maturity was occurring only in the two most dominant colonies. Despite the abundance of resources, outside of those breeding colonies, life was difficult—for female rats especially. At night, when the rats were active, any females not protected by a dominant male would be subject to almost constant sexual advances from packs of as many as a dozen males. "The exact number of mountings or attempted mountings per night is unknown," Calhoun observed, "but it must approach a thousand." The mounting behavior was relentless and debilitating:

> Practically all this activity occurs in the immediate vicinity of the female's home burrow. Only rarely does she get the opportunity to go to the Food Pen, and even then, males who paw and inspect her genital region or attempt to mount, give her little opportunity to eat or drink. There can be little doubt that this sequence of events produces a marked stress on the female's physiology.

Although pregnancies did occur in females subjected to this sort of intense attention, any resultant litters were usually abandoned, and almost none made it to adulthood.

The natural struggle for territorial control, which would normally have driven the rats farther afield to establish colonies elsewhere, was now diverted into senseless aggression. Over most of the pen, a state of violent instability held sway. Males that failed to mount a female would transfer that frustration into aggression, attacking young males who ventured near the food hoppers. Within a closed space, even a space the size of the Towson pen or a Baltimore city block, behaviors that had evolved to establish and grade dominance led not to social stability but only repeated conflict, disorder, and a near-permanent state of fight-or-flight. Although most adults survived amid this, the disruption to maternal behavior meant that infant mortality rose so high as to halt further population growth altogether.

Meanwhile, the broadly pyramidical social system, which saw one or two successful colonies perched atop a population largely struggling just to get by, was no unfortunate contingency but seemingly crucial to the rats' breeding cycle. The dominant colonies formed only by fighting their way to the top of the social hierarchy, and the dominant colonies performed almost all the reproductive work. Reproductive rates in the disorganized colonies fell below replacement levels; even when they were in mixed-sex groups, they either failed to breed or failed to raise enough pups to breeding age. Because dominant males controlled access to fertile females, only rats that were effective fighters passed on their genes, selecting for hardier strains and removing from the gene pool any traits that didn't cash out as powerful. Here was a self-organizing eugenics program that rendered weaker males functionally sterile, and in so doing acted to constrict the overall reproductive rate within a territory, reducing competition over available resources.

From a Darwinian perspective, the competitive behavior between the rats was an adaptation, evolved to ensure the species survived and thrived. Without the dominance gradient, the promulgation of weaker traits would erode their competitive edge. And so the consequences of a more egalitarian form of social organization were stark: "If rats failed to develop a class structured society," Calhoun contended, "they would not long survive as a species." Within a territory, the dominance gradient would always emerge: "The evolution of social behavior in the Norway rat has been such as to assure the development of a class-structured society." As Calhoun saw it, the inequalities between the colonies weren't a problem to be fixed, but a necessary condition for generational advancement.

———

TOWSON BORE the indelible imprint of the project's genesis in Baltimore, not only in its design but—unexpectedly—in its outcomes. The simulated city block simulated the city entire, with its powerful and socially mobile males in elegant dwellings on the edge of town, below which living standards followed a downward gradient through declining health and social mobility that bottomed out with homeless and marginalized outsiders living among abandoned structures. These connections were further facilitated by the rhetorical lens through which Calhoun presented his rodents. In the course of his observations, he stops talking about "rats" specifically, and instead refers to "individuals" or "young males" or "groups." Those are terms that permit slippage between species. All of this was not merely analogizing: *The Ecology and Sociology of the Norway Rat* also had a secondary agenda.

At first blush, the book's title reads as standard academic issue. It exhibits the unglamorous, utilitarian language of textbooks: *Basic Histology*, *Canine and Feline Cardiology*, *Molecular Biology of the Cell*, *Stress Fractures in Titanium* . . . *Ecology and Sociology of the Norway Rat* could hide on a bookshelf alongside those titles without standing out. But read it again, and there's something very peculiar. The usual technical term for the study of animal behavior is *ethology*. But Calhoun doesn't say ethology; he says "sociology." And sociology was exclusively the study of human social relationships. Calhoun was presenting something else: he's proposing an equivalence between the social behavior of rats and that of humans.

It is easy to scoff at this: Can we really talk about the "sociology" of rats? But as an ecologist, Calhoun's question comes from the other direction: Why have sociologists—who, after all, study the behavior of the mammal *Homo sapiens*—not paid more attention to the physiological effects of social interactions, both interpersonal and environmental? Walter Cannon's pioneering work in the trenches of the World War I had already anticipated some of this with the activity of the adrenal system in response to sudden fear or threat. Hans Selye had mapped Cannon's studies of acute shock onto the lower-level tensions of everyday life, onto how the cumulative effects of chronic stress built ultimately to exhaustion and ill health. Looking at how his rats' behaviors, social rank, and even reproductive fitness shaped and were in turn shaped by the structure and distribution of their burrows, Calhoun now sought to consider more closely the role played by space in social interactions, and how the steady erosion of bodily health through psychological stress might be understood, at least in part, as a consequence of the physical environment in which an animal finds itself. Just as Davis had urged that the study of the organism was inextricable from the study of its environmental interactions, so Calhoun sought to draw attention to the "physical determinants of social behavior." Nominally, he's talking about rats, but the lack of specifiers in the following passage is deliberate:

> The distribution and shape of objects or characteristics of the environment modify the frequency and pattern of contacts between individuals, which in turn are important variables in determining the size and stability of groups, as well as the stress experienced in relation to the frequency of making adjustments following contact.

In Towson, he had deliberately left the rats to themselves—the pen was relatively open, with just a few internal fences and the central food supply hopper to control the use of space. It is this middle section that Calhoun became most interested in. It was there, at the site where food and water and bedding were delivered, that the animals were forced into contact and confrontation. Most of the time, animals in separate colonies could keep out of one another's way. But when they came to the grain hopper or the water trough, the chances of meeting rats from a rival colony increased. Not only that, but the limited access points to the food supply also meant that a dominant rat could more easily defend that area, restricting others from using it. Calhoun wanted to examine that process more closely now, to shrink the enclosure down to just the central section, narrowing the experimental scope to focus only on where the unwanted interactions took place. Where the Towson enclosure had at least allowed the rats space to retreat and unwind after the exhausting business of venturing out for sustenance, new designs he began sketching for an enclosure that would focus more on the conditions that generated conflict and stress. He began to consider how one might manipulate the environment to make it more or less suitable for the animals that lived within it, and how those environmental parameters might differ between species. At the end of the experiment, he writes: “One question stands out: What are the configurations and spatial limitations requisite for an optimum state of existence for each species?”

Answering that question would have to wait. When the Rockefeller Foundation funding ran out in 1949, the Rodent Ecology Project was disbanded, and with it, the Towson enclosure. Calhoun’s contract with Johns Hopkins was due to expire in June. Although frustrated that he hadn’t been able to observe another breeding cycle, he feels a great sense of achievement, and of gratitude for the opportunity. “No one before me had ever had the opportunity to study such a complex social system from its inception,” he reflected, noting how the experience “had reset my course toward developing contacts with fields such as psychology, psychiatry, and sociology.” Be that as it may, the terminus was dispassionate and abrupt. A simple note records that in late May 1949, “all rats were captured and killed.” The dead rats were weighed and dissected, their scars counted, results tabulated. Among those still alive on the day of the cull was Male 49, who had maintained his undisputed dominance to the end. The records show that his body bore twenty-six wounds, and he had sired fifty-six surviving offspring—a third of the total population.

The hill behind the house was cleared, the observation tower dismantled and fences removed, the ground raked over. In 1958, the land either side of York Road was cut off from downtown Towson by the six-lane Baltimore Beltway. Today, the area is residential, affluent, and suburban. Private homes spaced wide across open lawns. A Carmelite monastery was established nearby in 1961, a golf course in the 1980s. The site of the enclosure is a corner on Belmore Road, where the blacktop bulges to form a parking circle. Some of the old growth woodland still stands, but there’s no trace of the quarter acre pen, nor the descendants of the Parson’s Island rats who, for twenty-seven months, had established their own empire on such an artificial island.

CHAPTER 6

MAXIMUM HUMAN PROTOPLASM

As Calhoun is packing up his things in Towson in late May of 1949, he happens upon a report from an awards ceremony in Newark that took place the previous week. The subject matter is population growth. Calhoun is intrigued. In an address to the American Chemical Society, plastics specialist Eugene Rochow proposes a thought experiment: What steps would be required to maximize the human population on planet Earth?

Rochow takes it as a given that this is a desirable goal. After all, as human life is "the most valuable thing on earth,"[1] then it follows that the more humans are alive, the more value there will be in the world. Rochow's approach to growing the planetary population is strictly practical: if human life is considered as a chemical system, then the problem is simply one of squaring the fixed needs of human physiology with the limited availability of suitable resources in the environment. The rate-limiting step on population growth is the supply of suitable resources. Enter the chemist, increasingly able to synthesize and produce on demand whatever complex molecules and compounds might be required.

Here was a drab and utilitarian future, where the planet is treated like an enormous strip mine from which to extract maximal calories. Sacrifices will have to be made to get the numbers up. With insufficient arable land to support their diet of "hot dogs and ice cream," the American people will need to utilize more plant proteins, augmented by cellulose from pulped tree trunks hydrolyzed to yield digestible sugars. Edible fats can even be derived from coal—a feat desperate German scientists had achieved during the war, yielding a sort of margarine dubbed "coal butter." A "satisfactory product," Rochow calls it.

Hampering basic food production on such a scale will be the carbon cycle. The "carbon problem" for Rochow is finding a sufficient supply of CO_2—in 1949, there simply wasn't enough of it in the atmosphere:

> Since much CO_2 is absorbed and combined by the alkaline oceans and deposited as a slime of calcium carbonate, we continually have less to deal with. We could resort to decomposing limestone, dolomite, and the other carbonate rocks while waiting for the sea bottoms to rise and replenish our supply, but the prospect of obtaining carbon only as CO_2 is rather discouraging.

Assuming we could find a way to increase atmospheric CO_2 in order to feed all those crops, Rochow lays out the expected human yield: "A US population of about one billion people could probably be fed, and a world population of 15 billions would not seem unreasonable."

The speech makes enough of a splash to be reported in *The New York Times* the following week in an article that described his "stimulating address" as "enough to make Malthusians in our midst shudder." Noting the extraordinary level of centralized control Rochow's form of utopia would require, the *Times* drily closes their editorial by suggesting that "Some Ph.D. ought to find out to what extent true democracy and the mechanization of life are compatible."[2]

Calhoun also finds himself disturbed by Rochow's proposals. But while the dreary prospect of living off of a diet of coal butter and refined cellulose pulp was grim enough, he is more troubled by the view that all that was required to sustain life was sufficient nutrition. His Towson rats did not face any problems with food supply, yet their numbers settled far below what their environment could have sustained. Nor was it calorific deficits that controlled the size of block populations in Baltimore. Reading the Rochow speech clarifies something for Calhoun: the obstacle to population growth isn't a dearth of calories.

Rochow's somewhat idiosyncratic plans for a massively expanded global population—and his choosing to frame the problem in terms of resource availability—came in response to the publication in the previous year of two hugely influential books: William Vogt's *Road to Survival* and Fairfield Osborn's *Our Plundered Planet.* Vogt and Osborn's readers would discover that all living things existed within a precariously thin layer, a film of organic matter smeared on the outer surface of a barren rock. Shattering the assumption of earth's boundless providence, these two books sparked in the wider public a new sense of our ecological vulnerability. To a significant extent, what became known as the environmental movement really began in that summer of 1948.

America had only recently emerged from the Dust Bowl years of the 1930s, when deep plowing of the Midwestern prairies stripped out the root structures holding the topsoil in place, allowing hundreds of millions of tons of arable land from Nebraska to New Mexico to simply blow away in vast black clouds. The sort of intensive farming Rochow had proposed as a means of feeding an ever-expanding population would only accelerate such erosion.

In the late 1940s, the world population stood at almost two and a half billion. Scientists estimated that if things carried on as they were, then by the end of the century that number might approach three billion. In fact, it reached the three-billion mark in 1960, and had passed six billion by the year 2000.[3] Overpopulation—that is, a population in excess of the carrying capacity of its environment—was detrimental to the biosphere, while competition for resources would eventually but inevitably cause further conflict.

In the coming decades, it would be this issue of unchecked population growth—rather than unchecked environmental degradation—that really captured the public imagination. Assuming the science was correct, the human population would eventually exceed the carrying capacity of the planet and collapse, returning the land to ecological harmony. Rochow's scenario imagined a future where the collapse never came, and the population simply grew larger and larger. To Calhoun, both outcomes seemed dire. But Rochow's scenario frightened him more. As Calhoun saw it:

> There is a great effort being made to increase the foodstuffs available to man. A derivative, if not its motivation, of this effort is the widely held opinion that we should produce more foodstuffs in order that more people can be born and survive. This demands that we maintain so long as possible an expanding population. There is no doubt that agriculturists and bio-chemists have made, and may continue to make, marked strides in accomplishing this objective. The question remains: What will be the impaction on the lives of individuals and the structure of groups as we progress toward this goal of maximum human protoplasm?[4]

Towson never had more than two hundred rats. Calhoun began to wonder: What was the upper limit? Assuming the food supply kept pace, what forfeit on the population for exceeding that?

Jack Calhoun wrote to Gene Rochow: Had he considered the psychological and social consequences of this level of population density? Where would everyone live, and more importantly, *how* would they live? Rochow replied that the "chief difficulty" would be feeding one billion Americans—it would require using nearly all of the available land for agriculture, which would in turn demand greater concentrations of population density within the cities. But, seemingly keen to record that he had taken the psychological effects of urban crowding into account, "I should like to point out," Rochow wrote, "that the high population density in large cities is in part alleviated by the purposeful isolation of many individuals."[5]

That "purposeful isolation" was troubling. When Calhoun had remarked that it ought to have been possible to accommodate five thousand rats in the Towson enclosure, it had been on the assumption that those rats were each housed in a standard cage, arrayed floor to ceiling as per the rat rooms Richter had built at Johns Hopkins, or in almost any scientific laboratory where rats were kept for research. The reason it was possible to house five thousand caged rats in an area that could support no more than two hundred loosed rats was the cages. The cages kept them apart. Isolated, each rat was spared the stress of competition, exempted from the natural urge to contest dominance for access to breeding females. Breeding in a laboratory setting was strictly controlled, involving the selective introduction of a single male to a single female in heat, and so female lab rats were liberated from the stress of relentless attention and unwanted mounting, while males need not fight one another for reproductive privileges.

Life for the lab rat seems civilized, orderly, idyllic, and the lab rats in response display exceptional complicity with their captors, apparently happy to be handled and to collaborate with whatever the laboratory requires of them. But the orderliness of the laboratory was achieved only by severely constricting the behavioral repertoire of the rats it housed, and further facilitated by the selectively bred docility of the lab rat itself. The population density achieved in the labs was possible only because the rats experienced a grossly etiolated social life. You could have five thousand rats in cages, or under two hundred roaming free. Rochow was advocating for the cages.

PURPOSEFUL ISOLATION isn't only an adaptation to high density, but apparently a positive lifestyle choice. An article in the Christmas edition of *House Beautiful* magazine in 1948 catches Jack's eye: "How to have a Private Estate on 105' by 103'." Perhaps he notices it because the dimensions are almost identical to the Towson enclosure. Showcased was the Tahquitz River Estates in Palm Springs, California, built by an up-and-coming real estate developer called Paul Trousdale. A few years later, Trousdale will buy up a swathe of wooded hillside north of the Sunset Strip in Los Angeles and shrewdly persuade the municipal authorities to incorporate the land within Beverly Hills. The Trousdale Estates neighborhood is now one of the most expensive addresses in America.

The selling point of Tahquitz River Estates is privacy: "Privacy is the keystone to good living,"[6] the article explained, and here was "a new kind of neighborhood where every homeowner gets 100% use of his own land." Across two pages of *House Beautiful*, a stylish line drawing showed six properties arranged in a grid, with the walls of the houses extending seamlessly outward to form integrated fences, effectively sealing each residence off from its neighbor. A caption explains: "The way to live privately is to face your house inward on your lot, as here. Then completely enclose the remainder of the lot with well-styled fences." Designed by architect Alan Sibel, the homes exemplify the midcentury Californian aesthetic: low and rectilinear, with plate glass walls shaded under wide eaves supported by pencil-thin pillars. Functional and orderly, they exhibit elegance without ornament. Each plot has its own pool. The Palm Springs estates were everything the cramped row houses of Baltimore were not: where a similar block in Pigtown might house twenty-six families, here there are only six. Each quarter-acre lot is its own domain: once you've pulled up onto your private driveway, you need never see your neighbor.

To Calhoun, they seem awful: "There are no alleys in the blocks and each house is completely surrounded by a high fence with no communication between adjoining lots. In seeking privacy, social isolation is achieved." The architecture embodies a rejection of community in favor of individualism: facing inward on your own lot, jealously enclosing one hundred percent of your own land. No children will be running over the neighbor's lawns here. Calhoun thinks of the rat pups growing up in the elongate burrows, how they emerged stunted and poorly socialized. The hermetic desert tract houses seem to him no better than the inner-city terraces, and perhaps worse: "I suspect that any housing technique that produces social isolation of developing children is every bit as deleterious as the excessive adjustments required by children and adults living in crowded tenement areas."[7]

As the Tahquitz River Estates was taking shape in the Coachella Valley, on the opposite side of the continent the first residents were moving into a very different sort of development. At the 1939 World's Fair in Flushing Meadows, New York, the Metropolitan Life Insurance company had displayed an intricate scale model of futuristic urban living: fifty-one massive, nearly identical buildings, each either eight or twelve stories high, spaced out over one hundred and twenty-nine acres and scheduled to accommodate as many as forty thousand residents.[8] Unlike the other utopian dioramas and mock-ups at the World's Fair, this one was real, or soon would be.

By the early 1930s, Metropolitan Life was by far the largest insurance company in America, providing services to a fifth of the population, and fully a third of city dwellers. The insecurity of Depression-era government bonds and a volatile stock market meant that corporations with significant capital reserves were in search of stable repositories for their wealth. Metropolitan Life had turned to investment in real estate, and they had an appetite for building big: the company's New York headquarters had been the world's tallest building when it was completed in 1909, and MetLife had subsequently financed the construction of the Empire State Building in

1930. Big housing was a logical progression. So, in 1939, construction began on a new development in the East Bronx. Situated between the Park Versailles and Westchester neighborhoods, the first Parkchester Apartments opened in 1941, and the development was fully occupied by 1943.

Robert Moses, an ambitious urban planner who had, since the early 1920s, steadily and skillfully expanded the remit of his official job as Parks Commissioner to become one of the most powerful figures in City Hall, and who would dominate New York's public building projects for almost half a century, looked at Parkchester with a mixture of awe and envy. His state budget, even aided by the federal government's residual New Deal funds, could never afford such a project. So Moses began to investigate and instigate alternative sources of capital. A series of tweaks to the legal regulations in the early 1940s opened a route for Metropolitan Life to divert their money into financing city projects in exchange for twenty-five years of tax exemption and a steady return on the investment. In case of objections or resistance from existing residents, the city would make available its "power of condemnation" to compel demolitions and evictions on MetLife's behalf.[9] That arrangement was significant. Where the relatively low-budget Baltimore Plan had relied on using novel legislation to incentivize an owner-led program of property maintenance, in New York, the city authorities had created legislation that allowed for private investment in public housing, enabling them to tap directly into commercial cash reservoirs to finance far more ambitious projects.

Stuyvesant Town was to occupy eighteen city blocks of the old Gas House district, providing 8,773 apartments across thirty-five high-rise towers and subsuming several churches, a few schools, and all the previously public roads therein. On the north side of 20th Street, a sister project, Peter Cooper Village, was built to a similar plan. Together, they formed a complex of fifty-six apartment buildings. It was trumpeted as "the first major effort made anywhere in the United States . . . to combine the powers of government and the resources of a great financial institution in a large-scale attack on the problem of neighborhood decay in cities."

Stuyvesant didn't look like the rest of New York City. Its towers were not aligned with the compass-point gridiron of Manhattan but arrayed instead around an oval park at the center of the development, a green space invisible outwith the crucible of buildings. It was as if the development had its back to New York. Critics called it a "walled city" and "a medieval enclave."[10] Like the compounds of the Tahquitz River Estates, albeit on a vast scale, Stuyvesant Town seemed to face in on itself, to fence in one hundred percent of its own lot.

Architectural critic Lewis Mumford clashed with Moses over the design, calling it a "nightmare" that embodied "the architecture of the police state."[11] Where Moses saw modern efficiency, Mumford saw a warehouse for storing humans, designed backward from the target of maximum occupancy: "Once the decision was made to house twenty-four thousand people on a site that should not be made to hold more than six thousand, all other faults followed almost automatically." Moses, in response, dismissed Mumford as an idealist, a "paper planner" with no real-world experience.[12]

As *Time* magazine reported on the spat,[13] Calhoun read about the dispute with interest. Although inclined to agree with Mumford's complaints, he conceded that "there was no rational basis dependent upon proof which backs up Mumford's contentions."[14] As it stood, there was simply no evidence that high population density and social isolation were bad for human development. And in the absence of any experimental data, it was essentially an argument about aesthetics: Mumford simply didn't like the look of it.

DURING THE two years Calhoun had been monitoring the Towson enclosure, he was still working with Davis and Emlen on the Rodent Ecology Project, spending weekdays among the row houses of downtown Baltimore. Being among the people by day, the rats by night, he began to see connections. As his attention alternated between the city and the pen, it left a sort of conceptual afterimage. He found himself thinking of the city in ecological terms and thinking of his rats in sociological terms, a convergence that would shape how he wrote up his findings in *The Ecology and Sociology of the Norway Rat.*

That latent sense that social science and natural science ought to inform one another had coalesced suddenly when he came across an article entitled "Concerning 'Social Physics,'" published in *Scientific American* in May 1948.[15] Its author, John Quincy Stewart, was a Princeton astrophysicist who wanted to explore the possibilities of talking about human societies in the same terms as physicists talk about molecules. In physics, the behavior of an individual particle might be random and unpredictable, but the behavior of a sufficiently large aggregation of such particles—in a liquid or a gas—exhibited obedience to physical laws, becoming predictable as such. Stewart now suggested that human populations could and should be studied in much the same way. Although societies were composed of free-thinking individuals, order and pattern might yet obtain at the aggregate level.

Catalyzing Stewart's "social physics" were the recent discoveries of George Zipf, a Harvard linguist who had been investigating a peculiar property of word distribution in texts. If you ranked all the words in a sufficiently large sample of text by how frequently they were used, a striking regularity emerged. The most commonly used word ("the") appeared twice as often as the second most commonly used word ("of"), three times as often as the third most commonly used word ("and"), four times as often as the fourth most commonly used word, and so on. The frequency of words followed a power law, whereby its rank was inversely proportional to its frequency. The phenomenon was remarkably consistent, and not just in English, but in every language he tested. To his astonishment, Zipf noticed the same pattern elsewhere. In 1950, New York had a population of 7.9 million. That was approximately twice the population of the second largest city, Chicago, which had 3.6 million, and three times the population of third largest city, Philadelphia, which had 2 million residents. And so it went on. The same pattern: a city's population ranking was inversely proportional to its size. And it wasn't just in America—it held for many countries, across a century of data. No one planned the cities that way; the pattern emerged spontaneously.

For Stewart, what would later become known as "Zipf's Law" carried the tantalizing promise that if "human behavior can be averaged" it might—just like gases and fluids—be seen to obey certain laws. Just as the water molecule does not know that it is part of a millpond or a whirlpool, any larger patterns or trends within a society would be invisible to the individual actors whose actions constituted them, becoming visible only through the sort of statistical analyses that so successfully described the motion of molecules *en masse*. The qualification for that visibility was that you had to view the societies from a sufficient distance, which required a very large sample—the sort of numbers Rochow envisioned, maybe more. A population of many millions—billions, even—might not be enough. Be that as it may, the accumulation of increasingly voluminous datasets about human societies—in the form of censuses, and by sundry government bureaus, in the activity of telephone switchboards, of vehicle usage, marketing research, tabulated sales figures, crime statistics, and spending habits—ought to allow for the tentative formation of testable hypotheses regarding any such laws.

Over just three flamboyantly speculative pages, Stewart spoke of the formation of cities in terms of "demographic gravitation," of population density in terms of concentrations of "human gas," and the influence of people on other people by analogy with electrical charge as "potential." The innate properties of the human atoms even offered resistance to the otherwise inexorable force of demographic gravitation: "Were it not for the expansive force of the human gas, representing the need of individuals for elbow room, the center-seeking force of gravitation would eventually pile everyone up at one place." When Stewart describes how every person has an influence on every other person, whose effect is inversely proportional to the intervening distance, Calhoun makes a note: "what an idea!"

In publishing "Concerning 'Social Physics,'" Stewart had given Calhoun tacit permission to transgress his own disciplinary borders. It felt as if here was a new language with which to speak about social behavior—not simply the growth curves and asymptotes of mathematically modelled populations, but fundamental principles governing the actions of individuals within those populations. Why aggregations occurred, why populations stabilized. Most enticingly for Calhoun: it offered the possibility of reconciling his insistence that any social group must be understood as being composed of individuals with his sense that larger forces acted upon those individuals as part of a group. In short: a conceptual bridge between individual choice and group behavior. Massed aggregations exhibited properties that were not visible at the level of the individual components. If there was a science of fluid dynamics for molecules, might there be for individuals a comparable science of population dynamics?

The idea of social physics helped Calhoun to see how he might connect his specific work on rats to more universal issues: how variables such as population density and the spatial arrangements of objects within an environment affected movements and social behavior. When Calhoun thought about social activity, it was increasingly in relation to physical activity: "Time and distance became impressed upon me as prime variables in any formulation of social behavior."[16]

The promise of lawlike patterns emerging from uncoordinated action had another implication. As Richter had intuited in 1920 when he first sat and watched his new rats: given the right frame of reference, movement was never random. Tracking the activity of rats with Davis and Emlen in Baltimore, and alone in Towson, had shown Calhoun that they didn't simply wander around freely but followed strictly the same routes to and from the same food sources and back to the same burrows. The location of a nesting site, the shape of the burrow, the particular trails they used all comprised "a physical mold in which the social matrix takes its form."[17] And that process was continuous across generations—young rats didn't build entirely new burrow systems, they reused the burrows they were born in, followed the same paths. The decisions made by earlier generations of rats were felt by future generations. There was, of course, a name for this process: it was culture.

> This alteration of the habits and social behavior of one generation by the activities of generations which precede it represents a cultural process, when culture is considered from a broad biological perspective. [. . .] There are striking similarities between the culture of man and that of some of the other vertebrates. Artefacts are constructed, learned patterns of social behavior are developed, and both are passed on to influence the life of later generations.

Davis had urged his team to treat the city as their laboratory, but Calhoun increasingly saw the apparatus of the laboratory in the architecture of the city, the same technologies of confinement. The city rats might be wild, but they weren't free.

"There, in Baltimore," he wrote, "rats lived in nearly closed cages, big boxes walled in by row houses. Furthermore, the enclosed space was divided into cells by high board fences."[18] The row houses formed an enclosure, a habitat designed by humans, within which the rats had built their own habitat, with its own paths and dwellings. Designing the interior fences of the Towson enclosure, Calhoun had tried to mimic the obstacles and barriers that separated rat colonies in the townhouse blocks. At Johns Hopkins, Richter had kept his rats in containers not much bigger than a shoebox, arrayed floor to ceiling in neat grids. When Calhoun thought of the inner-city backyards, they too seemed to him "like open shoe boxes"—"Lined up in two parallel rows and sealed by the connected houses on two sides of the block, these shoeboxes formed a larger, nearly closed box, slightly open at opposite ends by an intervening alley. Many blocks formed a grid."[19] The wild rats caged in Richter's Carnegie lab, the Parson's Island rats trapped inside his Towson pen, the city rats trapped within their block.

It occurred to Calhoun that nature, too, built fences, formed grids.

Just as the rats used the burrows their ancestors had dug, so the city came ready-made. It was not so much built as occupied: existing structures repurposed, new stories rising on the same old plots. Calhoun began to think of architecture as cultural inheritance. Long after Baltimore rescinded its official policy of racial segregation, the city blocks continued to exert inert control. This was why the architecture of Stuyvesant Town and Tahquitz River Estates mattered to him. Just as Poppleton's grid held a spatial legacy over the lives of Baltimore's current citizens, so any new development would itself become determinate, a dead hand reaching into the future, a mold into which the social matrix must grow.

In the immediate postwar years, the construction of new housing acquired special urgency. The American population had increased by fifteen percent during the 1940s to reach one hundred and fifty million by midcentury, a total almost exactly double that of 1900. Over a million immigrants, many driven out of Europe by the war, arrived in America during the decade; internally, over one and half million, mainly black Southerners, would move north to Detroit, Chicago, New York, Philadelphia, and Baltimore in what became known as the Second Great Migration. The overwhelming majority of both the internal migration and external immigration was into the cities. A little over a third of Americans lived in cities at the turn of the century; by 1950 almost two-thirds did. America was becoming an urban nation.

Stewart's concept of demographic gravitation, of populations clumping together into ever-denser aggregates, seemed consonant with the data. Cities were acting as attractors. The repeating towers of Stuyvesant Town would be repeated across the United States, at Pruitt-Igoe in St. Louis and Cabrini-Green in Chicago. Even as the cities became more crowded, tract housing was spreading outward into vast suburbs. On Long Island, builders

Levitt & Sons applied the principles of Fordian mass production to the construction of two thousand new homes, a process so efficient they were able to complete as many as thirty houses each day. New "Levittowns" would later spring up in Pennsylvania, New Jersey, and Maryland. Shoeboxes, lab cages, tract housing, and tower blocks became, for Calhoun, a blur of building projects: "This highrise, that school, suburbia, factories . . . quality of life, fulfillment, wino, mental hospital."

Against this backdrop, all began to cohere: how the promise of privacy tripped into social isolation, how increasing population density led to social strife. Calhoun stopped to wonder: "What kind of boxes are we building? For whom?"

CHAPTER 7

BAR HARBOR, WALTER REED

ALONG Maine's northern coast, where the granite and slate has been nibbled and scraped by retreating glaciers, the land shivers into thousands of islands, as if the continent itself were dissolving into the sea. The largest of these islands is Mount Desert, whose southern portion comprises Acadia National Park. Following several exceptionally dry months in 1947, wildfires devastated over seventeen thousand acres of forest surrounding the coastal town of Bar Harbor on the northeast of the island. Eight other towns and villages were engulfed, and sixteen people died. Also lost to the fire: the original premises of the Roscoe B. Jackson Memorial Laboratory, a biological research facility specializing in animal experimentation. Some ninety thousand mice were incinerated. Hamilton Station, a branch laboratory located several miles north of Bar Harbor in a cove on the Mount Desert Narrows, was untouched.

At Hamilton Station is the recently established Division of Behavior Studies, headed by John Paul Scott. Scott had been a student of W. C. Allee at Chicago, and—like Calhoun—retained Allee's emphasis on both the importance of cooperation and the heritability of behavior. While the Bar Harbor site was being rebuilt, Scott was engaged in the early stages of an extended research program on dogs. Although he called the project "A School for Dogs," the purpose was not obedience training but observing the natural expression of canine behavior. For thousands of years, dogs had lived among people. But as Scott saw it, our close relationship to dogs means that we actually know very little about them. Just as Richter had deviated from Watson's behaviorism to focus on "what animals do on their own, not on what they can be taught to do," so Scott wanted to see how dogs behaved when they weren't household pets. He wanted to escape our fond image of the dog as "a four-legged and childish human dressed up in a fur coat" and get to grips with a more basic question: "What kind of animal is a dog?"[1]

With his collaborator John L. Fuller, Scott would spend thirteen years studying the growth and development of over five hundred dogs at Hamilton Station, raising them in large open pens with minimal human interference. Although they performed problem-solving tests, which required contact with the scientists, for much of the study the dogs were furtively observed through small slit windows or over the compound's fences, with just their pack for company. Like Calhoun's Towson enclosure, Scott's School for Dogs was set up to contain the animal and conceal the observer, albeit for quite the opposite reason: where the rats were spooked by the presence of a scientist, the dogs were eager to meet them. Both reactions disrupted attempts to observe natural behaviors. Scott's efforts to stay hidden were only half successful: in photographs taken from observation posts above the compounds, the dogs are looking up to the camera, their tails all ablur.

Comparing the differential habits of several breeds and crosses, Scott and Fuller's investigations culminated in a landmark 1965 book, *Genetics and the Social Behavior of the Dog*, which remains an authoritative work on canine development. The wider aim of the School for Dogs was to investigate the genetic basis of behavior. Dogs were chosen because they displayed exceptionally high variability in appearance, and it was inferred a similarly high degree of behavioral differences might be found.

Studying dogs in a forest clearing overlooking the Atlantic doesn't seem especially contentious. But to talk, as Scott did, and Richter before him, about allowing an animal to "express its nature" assumed that each species had a nature to express. That is, a repertoire of behavioral characteristics particular to that animal and that arrived preloaded at birth. For the first three decades of the twentieth century, the idea that behavior, no less than hair color or height, was significantly determined by heredity in both humans and nonhuman animals had been widespread and relatively uncontroversial. But by 1945, when Scott first began planning his School for Dogs, the notion of innate characteristics had taken on a very different context.

Research at the Behavioral Studies Laboratory was predicated on both the heritability of behavior and the assumption that findings derived from animal studies would be broadly applicable to humans. The idea that the frequency of traits within a population could be controlled through selective breeding was a foundational premise of their research program. It was also the foundational premise of eugenics, and that generated a certain amount of hesitancy. Scott would later remember how most scientists expressed such "revulsion against the exaggerated claims of the early eugenicists and the racist doctrines of Nazi Germany" that many simply declared that "heredity had no effect upon human behavior." It was in response to this retreat from the stigma of eugenics that Scott's unit at Hamilton Station had been created. The founding director of the Jackson Lab, Clarence Cook Little, came to feel that important scientific questions were being neglected in deference to a political sensitivity. Little established the Behavioral Studies Laboratory and hired Scott to address directly the extent to which genetics influenced behavior.

Scott had organized a series of conferences in the late 1940s and early 1950s to bring together scientists working on the genetics of animal behavior. In 1948, the New York Zoological Society hosted a delegation on the "Study of Animal Societies," which Calhoun attended. Shortly after, Calhoun collected his thoughts on animal behavior in a short document he sent to Scott. He writes the principles in bullet points, as if composing a manifesto:

> The behavior of animals has a primary heredity basis.
>
> The basic hereditarily determined behavior patterns may be modified by cultural processes or other conditioning phenomena within the environment.
>
> The behavior of most animals is of such nature as to insure survival of the species (note that we do not say "survival of the individual".) through a given range of environmental conditions.

Following this, he adds a corollary: "Why do we study the problem of behavior?":

> To arrive at methods of controlling the development, the modification, and the expression of behavior in man, so that individuals may live in the social milieu of modern industrial civilization with a minimum of "emotional" stress within the individual, and with the minimum of conflict between individuals.
>
> To learn how to modify the environment of the earth so that man as an abundant species and the dominant species may have his needs fulfilled without upsetting the balance of nature as to eventually be detrimental to man.[2]

WHILE THE Calhouns were living in Towson, Edith had taken a job indexing scientific reports with the Child Study Clinic, an outpost of what soon became the National Institute of Mental Health (NIMH) at Bethesda, just outside Washington, DC. She learned that NIMH had money available for a small number of Special Fellowships, and, as the Rockefeller funding was soon to expire, suggested Jack apply. As it happened, NIMH already knew of Calhoun's work.

After he presented a seminar at Johns Hopkins in 1947 and arranged an accompanying tour of the Towson site, word spread of the man from Johns Hopkins who had built a "Rat-City" in the woods.[3] The US Army sent a film crew to shoot some footage of his captive rats for an instructional movie about pest control—Calhoun would later use the unedited film to make frame-by-frame analyses of fights and mating routines. The Baltimore *Evening Sun* reported on Calhoun's work twice: first in May 1948, when he was profiled as part of his work on the Rodent Ecology Project ("almost nothing is known about rats. What little is, is known principally by Dr. Calhoun"[4]), and again the following May to announce the termination of the project ("Scientific 'Rat City' Being Destroyed; Its King Had Harem And 56 Offspring").[5] Curt Richter had been fascinated by the enclosure, and—always keen on gadgetry—arranged the loan of some military-issue night-vision "snooperscopes" to study the rats after dark.[6] When Calhoun wrote to Richter, he called him the "godfather" of the project.[7] And, shortly before the Towson enclosure was shut down, Calhoun received an unexpected visit by three men from Bethesda.[8] They asked about where his research might go next. Calhoun spoke enthusiastically of his plans for larger enclosures, longer experimental cycles. The men said that the federal government was unlikely to fund such a program, but did tip him off that there was a large old dairy barn on the Rockville Pike, which currently served as a facility for National Institutes of Health lab animals. Calhoun made a note, thanked the men, and they left. He never did find out "why they came or how they happened to hear of this unusual study of mine." Perhaps they put in a good word back in Bethesda: Calhoun's application for a Special Fellowship was approved. Sponsored by NIMH, there was a place for him at the Jackson Lab in Maine, with Paul Scott's Behavioral Studies unit.

The job offer came as a great relief to Calhoun personally, as well as professionally. Just as the contract expired at Johns Hopkins, the Calhouns learned that Edith was pregnant. They would welcome their first child, a daughter, Catherine, the following March. Also heading for Maine that summer: a pet dog they adopted and named Dee-Dee. The dog was a stray Jack had found roaming the land near the Towson enclosure and brought back to the house during winter. It was an uncharacteristically sentimental gesture, although even here Jack was thinking like a scientist. Many years later, Calhoun's daughter would learn that "Dee-Dee" was not quite a pet name, but simply descriptive initials: D.D.—for "displaced dog."

BEFORE LEAVING for Maine, Calhoun had written to Scott with a long list of research topics he thought could be fruitfully explored—the duration of infant dependency, the cause of social aggregations and schisms, how animals modify their environment—but was encouraged to settle first on the study of home range, which, given the relatively short period allotted by the fellowship, Scott suggested would provide a suitable focus. To some extent, it would be an extension of the research Davis and Emlen had been doing on the home range of the Baltimore rats. But Calhoun's experience measuring the movement of individual rats within his Towson pen, and his newfound infatuation with social physics, inclined him to think more formally about the use of space within a home range, and of the invisible borders between adjacent ranges. In the absence of actual fences, how did different rodent populations distribute themselves in relation to one another? How were stable territories negotiated and maintained?

In Baltimore, the colonies of rats had been parceled up into neat blocks according to the layout of the city's grid. The roads had acted as fences helping to keep the populations apart, and social interaction had been limited to competition within the blocks. The Towson enclosure had purposefully imitated this with the perimeter fence. Calhoun now wanted to investigate how home range operated in a more open-plan environment. Inside the restocked premises at Bar Harbor were all manner of experimental specimens: primarily mice, but also rats, dogs, rabbits, hamsters, and even axolotls. Although Calhoun would carry out several experiments with the Jackson Lab's remaining stocks of inbred mice (those housed at Hamilton had survived the fires), he was first going to the woods, where he would be working with native species in the surrounding forests—deer mice, red-backed voles, the common shrew.

The job at Bar Harbor was a chance to return to his childhood passion of studying animals in the wild. Rather than inner-city row houses, he would be among aspen and speckled alder, yellow birch and red oak. He was assigned a research student, Alden Hinckley, a Harvard sophomore visiting Bar Harbor during summer break. Their assignment was to trap small mammals and calculate the species composition and their relative density to build up a picture of habitat use and biodiversity. In July 1950, they set to work laying traplines in the woods, consulting by mail with Emlen and Davis for advice on how widely the traps should be spaced. Calhoun had Hinckley set eight traplines over an eighty-acre area, four hundred and eighty traps in all.

Before the days of remote cameras, radio tracking, and GPS, "depletion trapping"—killing all the small animals in the trap zone—was a standard method for estimating the total population of a species within a territory. Once they knew how many animals were present, they could work out how much space each animal had to itself and build up an estimate of the home range for each species. So each day, Hinckley would walk through the forest, checking the traps and making the count. Not every animal would be caught immediately, but most would eventually, and so they expected to collect fewer animals each day as more and more mice, voles, and shrews were caught and population density within the trapping area declined. But instead, something most peculiar occurred.

For the first three days, captures declined as predicted. But on the fourth day, Hinckley reported a larger catch than on the first day. Calhoun was puzzled, assuming the anomaly was just a random fluctuation: "Well, I thought this was just due to the vagaries of chance or some unrecognized climatic factor." But Hinckley's count was higher again the next day, and then the next: "On every successive day his report was the same—more animals than yesterday." So it continued, until by the fifteenth and final day their haul of captured rodents was three times that recorded on day one of the operation. Where the trap counts should have declined toward zero, they were instead accelerating.

"All of this," Calhoun thought, "was very perturbing."[9]

Over the following months, he developed a plausible hypothesis to account for the phenomenon. Because any social contact is risky, potentially leading to conflict or avoidable competition over resources, each animal has a sort of buffer zone around itself, a bubble of personal space. By producing signals such as vocalizations or scent marks, two animals can avoid unwanted contact by detecting the presence of one another remotely (and alternatively, if in heat, can actively advertise for contact by laying scent trails and producing mating calls). This mediation occurs both within and between species, and it is mapped over dominance. So in the woods, the smaller shrews will avoid the larger deer mice, and the deer mice will avoid the even larger red-backed voles. Each individual agrees to stay within its own area. A pact of territoriality is tacitly established, and population density evenly distributed throughout the locale. Along with the availability of foodstuffs, this need to control social contact is one of the determinants of home range.

But Hinckley and Calhoun's trapping had destabilized this delicate equilibrium. As the population within the trapping zone was reduced, animals whose home range was on the periphery of the depopulated area detected a "signal void" at one side. They couldn't smell or hear any other voles or shrews, and as no animal wants its home range to abut that of its neighbor, they preferentially moved toward the area of lower population density. This led to migration into the now-depopulated trap zone, where they were subsequently caught. Because they had just vacated the home ranges along the periphery of the trap zone, this reduced the population density in an even wider ring, which enticed the animals living adjacent to the recently evacuated area to move into it. And so on. The effect rippled outward in a chain reaction, drawing recruits from farther and farther afield, like the upper surface of sand draining through an hourglass. The result was an ever-accelerating migration into the population vacuum. If the initial depopulating happened rapidly, Calhoun conjectured, a swarm could occur at the site of the initial evacuation. If he was correct, then the very act of rodent eradication might generate a secondary infestation far worse than before. It was as if someone had pulled the plug out of a basin—they were rushing uncontrollably toward the center, all piling into the same small space. It reminded Calhoun of Stewart's idea that cities exerted "demographic gravitation," drawing citizens into ever-denser aggregations. Calhoun called it an *induced invasion*.

Here, perhaps, was a plausible mechanism to account for historically recorded rodent plagues—most iconically in Nordic lemmings, which were said to swarm together and suicidally hurl themselves into the sea, but also the plagues of mice that periodically destroyed grain stores in Australia and vole plagues in continental Europe.[10] In 1942, British ecologist Charles Elton had collected historical accounts of rodent plagues into a book called *Voles, Mice, and Lemmings* and sought to find a pattern in the boom-and-bust cycles.[11] In the arctic circle, lemming migrations were well-known. During plague years, the lemmings would emerge from the woods in vast numbers, moving always downhill, and so inevitably into the sea. No one had been counting lemmings, but the fur traders of the Hudson Bay Company had been counting pelts of lynx and foxes. Elton reckoned that lemming gluts during mass-migration years ought to boost the numbers of animals further up the food chain. Scouring the records, he discovered the haul of lynx pelts fluctuated in regular ten-year cycles, and inferred lemming swarms were the cause.[12] But the issue of why the rodents made their massed migrations simultaneously, and why they were seemingly compelled to continue moving long after they had discovered new pastures, remained a mystery. Elton tentatively suggested the pattern might be linked to sunspot cycles, but these averaged at eleven years—a periodicity that was close, but not close enough. Calhoun wondered if the phenomenon he had triggered in the Maine woods might provide an experimental opportunity: if the migration behavior could be induced at will, it could be studied.

While working on the Rodent Ecology Project at Johns Hopkins, Calhoun had been corresponding with field biologists across the country to collect data on rodent population cycles. The project was called the North American Census of Small Mammals and involved at its peak almost two hundred ecologists surveying local rodent species and posting their findings to the Rodent Ecology Project at Johns Hopkins, where Calhoun was responsible for collating the results in an annual report.[13] He sent out feelers to see if anyone else had witnessed a similar result, encouraging his network of ecologists to recreate the trapping experiment. However, initial interest in these induced invasions would come from outside ecology altogether. David McKenzie Rioch, a neuropsychiatric specialist at the Army Medical Service, also heard about the effect. To Rioch, who worked on the behavior of soldiers under stress, what Calhoun was describing looked a lot like troop panic.

Panic was a problem for the army. They recognized that stress was a perfectly reasonable reaction to a combat situation, but along with the "normal psychosomatic reactions" to severe stress—faintness, giddiness, nausea, and diarrhea—there were yet the "abnormal reactions": "grossly incapacitating shaking and tremors," "conversion hysteria; pseudopsychotic reactions," and "pathologic fear such as panic."[14] If everyday stress was a slow release of adrenaline, panic was its torrent, the desperate apotheosis of fight-or-flight, involving "temporary major disorganization of thinking and control of fear."[15]

In a sense, the military had been practicing behavioral conditioning long before it had a name. Psychologists formalized what every sergeant major already knew: drills and boot camps were a means of instilling automatic and predictable responses from recruits, and military discipline aimed to overwrite natural instincts with a regimental protocol. From the army's perspective, panic behavior was "the opposite of regimental behavior." Where the regiment "bound the crowd together into mass action, mass attention, mass strength" to control fear, panic was "the antithesis of crowd behavior." "A panicked crowd," one guidebook says, "is a disorganized herd." And a disorganized herd was exactly what Rioch saw in Calhoun's induced invasion. So in 1951, with his NIMH Fellowship in Maine expiring, Calhoun was invited back to Maryland to take up a post with the US Army.

THE ARMY Medical Center Research and Graduate School was housed in a suite of buildings adjacent to—but administratively separate from—the civilian National Institutes of Health in Bethesda. When Walter Cannon had been posted to France during World War I, the principal focus of the army's medical division had been on repairing broken bodies and maintaining the physical health of combat troops. The research agenda of the Army Medical Center had been focused on infectious disease, dentistry, and physiotherapy. But by the end of World War II, in which almost a quarter of all battlefield casualties had been recorded as psychiatric, the importance of mental well-being to overall fitness was impossible to ignore. Here, the concept of "stress" provided a bridge between mental and physical health. Couched in the language of bodily ailments, stress allowed the military to talk of psychological problems in robustly physiological terms. Selye's "general adaptation syndrome" had been the subject of a new Army information film and, among the research areas listed in the 1949 Army Medical Center Research and Graduate School prospectus, the only item explicitly addressed under mental health is "Effects of Chronic Stress Upon Resistance."[16] It was an indication that what had once been loosely gathered under the euphemistic banner of "troop morale" would now be examined more systematically.

Before the war, it had been widely assumed that mental breakdown during combat occurred in individuals already predisposed to mental illness. As a 1943 textbook on *War Neuroses* by psychiatrists Roy Grinker and John Spiegel put it: "Schizoid characters, who have managed to get along fairly well in stable conditions of civilian life, are apt to develop paranoid reactions when involved in the intensely hostile situation of the battlefield."[17] Combat was merely a catalyst aggravating latent problems. But as the war rolled on, enough reports had accumulated of even "extremely reliable men and officers"[18] suffering from acute mental trauma that there could be little doubt that anyone, even "individuals with a minimal predisposition to mental illness,"[19] were liable to be broken by the stresses of battle. By 1944, Grinker and Spiegel had revised their position, declaring categorically that: "War neuroses are caused by war; anyone, no matter how strong or stable, may develop a war neurosis."[20] By the late 1950s, a conference of Army psychologists was meeting annually to share their findings and discuss future projects.[21]

Crucial to this psychological turn had been David Rioch, director of the Army's Neuropsychiatry Division. Rioch had been lobbying the military to conduct more work on behavior for some time, and in 1951, as the Army Medical Center was reorganizing ahead of becoming the new Walter Reed Center, a dedicated Department of Experimental Psychology was established.

Rioch hired Captain Joe Brady to lead the unit. Just twenty-nine years old, Brady had already served as an infantry platoon leader in Europe and, when hostilities ceased, was posted to a former Luftwaffe hospital in Germany where—despite having no clinical experience—he was appointed Chief Clinical Psychologist. On his return to the US, Brady completed a Ph.D. at the University of Chicago, using rats in a behaviorist research program to investigate how stress and anxiety would respond to electroconvulsive interventions. He later undertook pioneering work on how unwanted side effects of drugs such as *reserpine*—officially indicated for high blood pressure—might have useful anti-anxiety and sedative properties. Working with monkeys and rats, Brady would later play a crucial role in establishing the field of psychopharmacology. On account of his success with managing anxiety in rhesus macaques, when NASA wanted to put a live animal into its Mercury rockets, Wernher Von Braun personally sought out Brady to train and assess the suitability of the space monkeys for orbital flight. At the point when Rioch approached him, Brady was stuck with a dead-end assignment carrying out Rorschach tests on inpatients at Walter Reed Army Hospital. Offered the chance to pursue psychology research, Brady was delighted to discover the army might actually "have a place for somebody who plugged rats into light circuits!"[22]

So Calhoun's research on rat behavior initially seemed to intersect well with the open-ended experimental approach Rioch and Brady were cultivating at Walter Reed. Rioch asked Calhoun to prepare a brief description of the type of work he might pursue on behalf of the army. Calhoun's plans look very similar to those he had been formulating toward the end of his stint at Johns Hopkins—examining how different environments facilitate or inhibit communication, group dynamics, and social order in rat colonies—but with an additional focus now on artificially stimulating panic reactions and mass migrations, per the army's interest. He submitted what was becoming a characteristically ambitious research program—seven pages of typed manuscript. Reviewing Calhoun's proposal, Brady noted: "John, If you put much more in here you will have Rioch's job."[23]

Calhoun worked alongside Brady designing enclosures to increase or reduce anxiety in lab animals. Brady was principally using operant conditioning chambers, also called Skinner boxes, which were metal crates hooked up to a measuring apparatus. Inside each, a solitary animal—a rhesus macaque, a rat—with a task to perform, or subjected to precisely calibrated stimuli from which precisely calibrated responses might be obtained. These were one-dimensional studies, and deliberately so. They were designed to isolate the attribute being examined, and minimize all other variables, eliminating the noise from the signal. Brady's work seemed to carry the military ethos of regimented control and discipline through to the treatment of the lab animals. Solitary confinement in small metal crates allowed for precise and predictable results, a standardized lab rat in a standardized container. When the lever was pressed, a record was automatically generated, and the data accumulated on spools of tape. Rat and box together formed a unit, a machine for the production of reliable and replicable results.

Calhoun wrote that the metal boxes made him "squeamish,"[24] and privately began to question the value of such experiments—particularly for behavioral research. His objection wasn't simply compassionate, although it was also this. Rather, it was that he grew skeptical of the usefulness of results produced by the conditioning chambers. As with the mathematical population models that Davis had objected to at the very start of the Rodent Ecology Project, the experiments seemed designed to confirm the hypotheses they were supposed to be testing. Will a rat administered an electric shock learn to press a button to stop the shock? Even if such a system told you what you wanted to know, it could *only* tell you what you wanted to know. It seemed that there was nothing else to learn, that no unexpected results were likely to emerge. Precision was purchased at the expense of discovery.

Laboratory studies involved a trade-off between observability and the expression of natural behaviors. In the field, too much went unseen, and there were too many uncontrolled variables. Calhoun acknowledged that. But in these metal crates, the artificiality of the environment, and the impossibility of social interaction, skewed the results too far. Rats were social creatures; their behavior changed in response to the activities of other rats. Calhoun knew from Towson that their physiological growth, immune system, and fertility varied according to their dominance rank. How to usefully measure anxiety or stress in an animal already stressed and anxious, and unable to exhibit the full range of its reactions? Measuring behavioral deviations from such a baseline seemed pointless.

Brady used Skinner boxes because he was employed in a Department of Experimental Psychology, and that's what experimental psychologists did. For Calhoun, it felt as if disciplinary conventions were taking precedence over scientific enquiry. Brady didn't think like an ecologist. To him, the rats were a component in a machine, recruits to be regimented.

The ecologist worked in the field, the psychologist in the lab. Calhoun thought there was a compromise between the two, a new type of laboratory. He made plans for building seminaturalistic enclosures, of a size and design that took account of the animal's innate behavioral repertoire. He would begin to call them *compound studies*. Such a space would permit the scientist to covertly observe the animals, generating and collecting data without intrusive interventions. Just as with his Towson enclosure, or Scott's School for Dogs, the compound study allowed the organisms to live in an environment that permitted greater variability of behavior and emotional reactions: "From animals reared and studied in compounds it is possible to secure a much wider range of types of known origin than can result from rearing under the more stereotyped caged conditions."

Since Towson, Calhoun had been drawing designs for enclosures that would allow rodents to express their natural behaviors and still be visible for scientific study. Circular pens divided into wedges of varying sizes, octagonal pens, hexagonal pens, high-rise pens with multiple floors, pens with mazelike barriers that would inhibit accidental collision, others with a central feeding area that would compel cooperation. Pens that look like Stuyvesant Town, pens that look like the Tahquitz River Estates. One proposed enclosure for studying "population gradients" was eight hundred feet long and five hundred feet wide. Another was a vast arena that covered some one hundred and twenty thousand square feet, divided into twelve radial segments surrounding a central observation tower. He anticipated this part of the experiment could run for a decade. [25]

Urged by Rioch to focus on achievable results, Calhoun tried to create laboratory studies that would meet the army's expectations. He glumly compiled a list of "Problems Capable of Completion in one Year." He built a long, single-track enclosure—thinking of his time in Baltimore, he called it an "alley"—through which a rat could only run back and forth. It was a compromise between the metal crates Brady was using and the type of enclosures he sketched out in his notes, something that allowed the expression of at least some natural behavior while isolating one variable. If he thought it might satisfy his employers, it didn't satisfy him as an ecologist: the experimental setup was too formal to produce unexpected results, it didn't simulate the dynamic complexity of a natural habitat.

Meanwhile, despite Calhoun's having been hired by the army to investigate induced invasions, the facilities at Walter Reed weren't suitable for recreating the swarming effects he and Hinckley had observed in Maine. Instead, an ecologist from upstate New York, William Webb, part of the Census of Small Mammals network that had been corresponding with Calhoun during this period, was subcontracted to recreate the trapping experiment on a larger scale. Working in the Huntington Forest reserve near Newcomb, New York, Webb expanded the operation. In 1951, he laid a ring of more than seven hundred traps, one every five feet, enclosing an area of almost twenty-three acres. Within this perimeter were further traps, over nine hundred in all. The initial operation ran for eighty consecutive days. Over the next three years, the experiment was repeated on different sites. As in Maine, typically, each trapping brought in a haul of deer mice, red-backed voles, and shrews.

With the ecological fieldwork largely delegated to Webb in upstate New York, Calhoun was free to spend much of his time at Walter Reed, writing up the results of his Towson experiment. Rioch recognized Calhoun's talent but worried that his attention was too diffuse. He encouraged Calhoun to get his notes into shape, to focus on definite questions that were likely to yield definite answers. Against the military methodology of Brady's studies, Calhoun's work looked decidedly civilian: untidy, open-ended, a mishmash of disciplinary borrowings and fragments. He accumulated great reams of data but published very little. Now, under Rioch's guidance, Calhoun began a more systematic approach to his output.

Collating the results from Webb's Huntington Forest operations and his own work from Bar Harbor, he was looking to understand the movement of rodents schematically across time and distance, seeking the sort of lawlike principles suggested by John Stewart's idea of "social physics." He began working with a mathematician, James Casby, an MIT graduate who joined the Neuropsychiatry Division as an electrical engineer. With Casby's help, Calhoun was able to use data collected from his own studies and those of the Census network to formally describe the use of space within a home range. He increasingly leaned toward the language of physics: "We began thinking of mice, or any animal for that matter, as molecules moving in a random fashion from and back to some fixed point, the home site."[26] The proximate outcome of this collaboration would be a formula-dense paper on home range and density in small mammals, but as he was spending much of his time writing up the results of his Towson experiment, the trace of Casby's mathematical influence would later emerge in the many equations and formulas that Calhoun included to describe rodent behavior in *The Ecology and Sociology of the Norway Rat.*

Although Calhoun and Webb published a couple of short reports in *Science*,[27] as the results from the Huntington Forest studies came in, the initial similarities Rioch had seen between Calhoun's induced invasions and troop panic were ebbing away. Webb's results showed an increase of trapped animals over time, suggesting sequential waves of immigration to the area roughly every two weeks, but he struggled to replicate the dramatic results that Hinckley and Calhoun had produced at Bar Harbor. By July 1953, Calhoun wrote to Webb to regret that "the efforts to produce a mass movement have not really been successful."[28]

There was another puzzle. At first, the forest studies only caught deer mice and red-backed voles. Later, they caught shrews—and in ever-greater numbers. After several months of continuous trapping, the haul of shrews was double that of the red-backed voles.[29] The shrews, threatened by the presence of nearby mice and voles, were usually subterranean, travelling in shallow burrows under the leaf mold, and consequently rarely caught in the surface traps. But after most or all of the more dominant voles and mice had been trapped and evacuated, shrews would venture back up to the surface and be caught in the traps. The behavioral repertoire of the shrews changed depending on the composition of local species.

The behavior of the shrews—becoming evasive in the presence of competitors—offered insight into a problem in ecology concerning the "competitive exclusion" principle. This states that if two species are doing the same thing in the same geographical territory, any slight advantage in favor of one of them ought to eventually drive the other out. There's only so much room in an ecological niche. Because the two species wouldn't be perfectly equal, the more successful species would ultimately displace entirely the less successful species. This, indeed, was the very process by which the Norway rat had largely displaced the black rat, first in Europe, then in America. Darwin himself had noted as much: "How frequently we hear of one species of rat taking the place of another under the most different climates!"[30] Nature lacks antitrust legislation, and, as with businesses in an unregulated market, niche occupation tends toward a monopoly.

If the principle was correct, there ought not to be two species doing the same thing in the same place. The coexistence of three species—mice, voles, and shrews—in the same place, doing the same sort of things, seemed to violate the competitive exclusion principle. But the shrews had developed novel behaviors—travelling almost exclusively underground, which allowed their territories to overlap with the mice and voles. If the area was considered in three dimensions, the shrews were occupying a lower depth of the same area.

As promising as the notion of an induced invasion might have been, the hypothesis did not hold up. When he reexamined the data in 1961, Calhoun came to a different conclusion: "I was completely *wrong* in the interpretation that the tremendous increase in catch Webb & I found within 15–30 days of removal trapping represented an invasion."[31] Instead of a rapid acceleration as it had first appeared, the increase in catch simply recorded an enlargement of the home range of subordinate species in the same area. With the dominant specimens removed, the more timid creatures had emerged and begun travelling freely on the surface, and in a larger range, where they were subsequently caught in traps their subterranean routes had previously avoided. The apparent "invasion" was really just a redistribution: a homeostasis of the habitat.

But the finding was still significant: The less-dominant species didn't inherently have a smaller home range but settled for a smaller domain in order to fit in the gaps between the home ranges of the more dominant species. When the dominant species was removed, the subordinate species expanded their wanderings to a larger zone. Here, finally, was an answer to the question Scott had first suggested Calhoun work on. And the dogmatic stance Calhoun had taken on heredity during his first season with Scott at Hamilton Station was softening, becoming more nuanced. The behavioral repertoire of an animal might be innate, but it was flexible; it changed according to the environment. What it meant to behave like a shrew was a function of having red-backed voles in the same habitat.

THESE ARE all interesting findings for ecologists, but they have little bearing on the issues the army had wanted to explore. In June 1953, Calhoun writes to Webb to warn him that Rioch "took a rather dim view" of their most recent round of research, and when Webb replies a few weeks later, it is to report that he has been told the current military grant will be "terminal."[32]

Meanwhile, seeking to extend his own contract with the army, Calhoun's attempts to initiate more work on social dynamics at Walter Reed are running into unexpected problems—although this time, they have nothing to do with the science and everything to do with politics. When he submits work to his superiors, it is being returned with more and more edits. An internal memorandum he writes in 1953 is titled "The theoretical framework from which I approach problems of social behavior." Except the word "social" has been redacted and replaced with "group." Further down the document, Calhoun has written: "Although the problem is stated in general biological terms applicable to any *social* species, it is one which is becoming particularly pertinent on the human level as the population becomes both more dense and more urban." In the margin, the sentence has been marked *omit*. Throughout the document, a pattern emerges: " . . . these principles will form a 'preventative medicine' of *social* behavior . . ."—*omit*; " . . . *social* instability of groups and *social* malfunction of the individual . . ."—*omit*.

For some time, Calhoun's intellectual interests have been widening out from ecology to encompass concerns about architecture, sociology, and psychology. And his role within the military has obliged him to make explicit the similarities between mice and men—to use his rodent studies as the basis for understanding human behavior. Not only does this demand a conceptual equivalence between social behavior in different species, it also naturalizes that social behavior. It makes sociality a background condition, assumes that humans are fundamentally and essentially social beings. However uncontroversial that may sound now, in a United States government facility in 1953, saying people were essentially *social* was a mere syllable away from saying they were essentially *socialist*.

This is the year Wisconsin senator Joseph McCarthy turns his anticommunist spotlight on the US Military. Under the gaze of the Senate Permanent Subcommittee on Investigations, Calhoun's research shows up a little too red. Fearful of attracting the Senate Committee's attentions, the colonel overseeing their divisional budget cancels funding for the research on social use of space. As Calhoun pleads his case, the colonel looks up at him: "What will McCarthy say?"[33]

His contract with the army will not be renewed.

Losing his job comes as a hard blow. Personally, he feels great warmth toward both Rioch and Brady, and admires both men. Calhoun will later say of Rioch that he "sheltered my searchings," and in Calhoun's promotion application Rioch describes him as "a terribly first-rate person"[34] and wishes that they could have kept him at Walter Reed. But even without the threat of McCarthy's witch hunt, Rioch knows that Calhoun's roving, open-ended approach to research is not a good fit for the results-oriented military system.

Jack Calhoun is worried that he might not be a good fit anywhere. The termination of his contract has occurred at a most unfortunate time: the Calhouns have just purchased a house on a leafy residential street in nearby Kensington, a couple of miles from Walter Reed, and Edith is pregnant again. At thirty-six years old, with a young family to support and a second child on the way, he is facing unemployment. Feeling sorry for him, Rioch agrees to keep Calhoun on the army payroll for as long as he can, but the situation is precarious.

His experience at the military has been disappointing: the failure to satisfactorily explain panic phenomenon, the vertical hierarchies of the command structure, his frustrated ambitions. At Bar Harbor, there was open space, natural habitats, an emphasis on social behavior and the expression of innate instincts. Walter Reed involved small cages, machines in which the animal was a solitary and malleable component. The army seems to prefer a more regimented version of experimentation focused on controlling rather than observing behavior—more Watson than Richter. Calhoun thinks of Joe Brady's "automated labs with rats and monkeys in isolated boxes pressing their levers and spewing out endless tapes of cumulative records."[35] When, contrastingly, he remembers his time in Maine, and "the mice and shrews in the field and forest, of their perils and freedoms," he is thinking also of his own intellectual freedom, and the perils of striking out alone.

Jack Calhoun is beginning to suffer the consequences of his disciplinary wandering. In January 1954, he writes to Paul Scott back in Bar Harbor with a sense of pessimistic resignation about his professional future: "I don't put much stock in the likelihood of securing a position in the wildlife field. Even getting back into a zoology department of a university poses quite a problem."[36] Scott is still at Hamilton Station, where he will continue running his School for Dogs well into the 1960s. In upstate New York, William Webb returns to his ecology work at Syracuse University. John Christian, who had spent the war on a torpedo boat, has been employed at the Naval Medical Research Institute where he has been appointed head of the Animal Laboratories. He is also finally completing his doctoral studies at Johns Hopkins. Supervising Christian's dissertation is David Davis. John Emlen is away in Europe on a Guggenheim Scholarship. Alden Hinckley is now at Berkeley, on route to a Ph.D. in entomology from the University of Hawaii. Hinckley will spend much of his career in the South Pacific working for the Environmental Protection Agency, helping farmers control tropical insects.

Calhoun is soon to be unemployed. He has a series of unfinished projects, a sheaf of fantastical-looking designs for enormous rodent enclosures, a new mortgage, and now a new baby—their second daughter, Cheryl, is born in April 1954. Disciplinary identity is another sort of box: you are a psychologist, or a sociologist, or a physicist. Jack's interests are liminal, interstitial: he is an ecologist whose professional habitat is the city, a naturalist more interested in simulated environments than field studies. Fascinated by sociality, he is horrified by crowds. While his peers and former colleagues seem to have stayed within their lanes and flourished, Calhoun wonders if his enthusiasm for connecting his work to other disciplines has been a terrible mistake, if he has spread himself too thin, and if by trying to straddle disciplines he has fallen into none.

CHAPTER 8

CASEY'S BARN

EUGENE Casey became a rich man in the Great Depression. During the 1930s, the building company he had founded while still a Georgetown law student secured a run of lucrative New Deal contracts. A generous donor to Franklin Roosevelt, Casey was soon rewarded with a political role: Director of the Farm Credit Administration, where he allocated government loans for victims of the dust bowl. The national housing shortage was a boon for builders such as Casey, and as the nation got back on its feet, the profits accrued. He invested in real estate, buying up swathes of land in Montgomery County, Maryland, including two farms of his own. He called them "The New Deal Farms," in tribute to the funds that had paid for them. And in 1938, where the old Shady Grove Road crossed what was then US Route 240 in the hinterland between Rockville and Gaithersburg, he built himself a barn. Visible from the highway, it was deliberately picturesque: three gambrel roofs in antique Dutch style, constructed with lumber from his own estates, the barn became a local landmark. During the 1940 presidential election campaign, Casey had the gables repainted as a three-story signboard endorsing Roosevelt for a third term of office. In return for the fundraising, Roosevelt appointed Casey as a special advisor on agriculture.

By the mid-1940s the towns of Gaithersburg to the north and Rockville to the south were encroaching on the New Deal Farms. Casey sold the cows, built low-cost housing on the pastures, and gifted lease of the barn to the National Institutes of Health, which needed a place to keep the rhesus monkeys they were using for trials of Jonas Salk's polio vaccine.[1]

By the early 1950s, with the animal phase complete and the vaccine testing moving onto human subjects, operations at the barn were winding down. In 1953, NIH opened dedicated animal buildings at their Bethesda campus, and by mid-1954, Casey's barn stood empty.[2]

BACK IN Maine, Paul Scott had told Jackson Lab director Clarence Little of Calhoun's situation. Little was fond of Jack Calhoun, and especially liked his work on spontaneous population regulation in rats. Little worried a lot about population growth. In 1921, he had been a founding director of the American Birth Control League. As a geneticist, Little was concerned with selective breeding, and his focus was on the management of heritable conditions and congenital disorders through targeted contraception. In theory, that was intended to improve societal health and reduce the burden on public resources. In practice, it meant providing access to contraception shaded into enforced sterilization.

In 1921, the same year the American Birth Control League was formed, Little became president of the University of Maine, where he established a research outpost at Bar Harbor, on Mount Desert Island. In 1929, he accepted the presidency of the American Eugenics Association. Even before the state-imposed sterilization programs of Nazi Germany, his advocacy of this version of "birth control" was a controversial stance. In 1939, the

American Birth Control League merged with the Birth Control Council of America, and in 1942—concerned that the semantic ambiguity of "birth control" unhelpfully elided the difference between women who didn't want children and women who shouldn't have children—it rebranded itself as the Planned Parenthood Federation of America.

By 1954, Little was on the board of an organization called the Population Reference Bureau. The PRB was a nonprofit based in DC that described itself as a "clearinghouse for population information."[3] Although the PRB strove to publish data on demographic trends without any political steer, the tilt toward population control was implicit. If the PRB had any coherent agenda, it certainly wasn't aimed at *increasing* the national birth rate.

The director of the PRB was Robert Cook, to whom Little had previously introduced Calhoun while he was at the Jackson Lab. Little now spoke to Cook. Bob Cook spoke to Alan Gregg, outgoing vice president of the Rockefeller Foundation, and arranged for him to meet with Calhoun. Over a long lunch at the Cosmos Club in Washington, Calhoun told Gregg of his ambitious research plans, that he wanted to study multiple generations of rats over many years. Funding bodies were wary of committing to large-scale, long-term projects, preferring to distribute smaller grants to a greater number of recipients.

Gregg felt this was all wrong. Good science took a long time, and while short-term grants were less risky, they were also less likely to yield substantive results. It was Gregg who had signed off on thirteen years of funding for Scott's School for Dogs. He had frequently clashed with the Rockefeller Foundation over their reluctance to support long-term projects, complaining that if the Foundation offered only short-term grants it risked becoming "the largest distributor of chicken feed in the country."[4] And so, rather than balk at Calhoun's ideas, he was impressed. On this occasion, Gregg didn't have funding, but he did have influence.

As Calhoun remembered it, "Dr. Gregg said in effect, 'Just leave it to me, I will arrange for you to be taken on at NIMH.'"[5] Gregg spoke to David

Shakow, recently appointed chief of the new Laboratory of Psychology at NIMH. Shakow, looking to build a team for his lab, agreed to take on Calhoun. Ricoh would continue to pay his salary from the army budget until the appointment officially began in June 1955.

At NIMH, which was located just the other side of the Rockville Pike from Walter Reed, Calhoun showed Shakow all the plans he had been making for environmental manipulation—detailed designs for exotic enclosures, hexagonal pens, mazelike warrens, rooms designed to increase social friction, rooms designed to increase cooperation. Calhoun already knew about the property NIMH were using as an animal house up past Rockville: the three men from Bethesda had mentioned it during their visit to Towson six years earlier. He suggested to Shakow that if the owner were willing to extend the lease, the barn might make a suitable base for his studies. Still frustrated that his experiment at Towson had been cut short before he could observe the behavior of the rats there into successive generations, he insisted the new experiments must be allowed to run for at least five years. NIMH agreed. Shakow asked Calhoun to submit a proposal.

REMODELLING OF the barn takes place over two years. Casey oversees the work, and pays for it all. As the project nears completion, Calhoun writes him a letter of gratitude:

> It is difficult for me to convey to you the real uniqueness of your contribution to medical progress. All recognize the importance of disease organisms and drug therapy. In these areas it is relatively easy to obtain support for personnel, equipment, and laboratories. Much more recently there has developed a cognizance that a person's mental well-being is a component aspect of his over-all health. Here effort and

> research monies have concentrated on the psychotherapy and drug therapy of those already mentally ill. Development of a program of "preventative medicine" relating to mental health is just on the horizon. Part of this horizon is a conviction and little more than that, that the physical surroundings a person lives in affects his general state of happiness and the satisfactoriness of his relationship to his family and neighbors . . . Exactly where this study will lead I cannot say, but just as surely as studies with rats led to the discovery of vitamins and all this has meant for nutrition, there is equal likelihood that our studies of the relationship of a rat's environment to its physiology and "mental health" will lead to insight of value to human problems.

WHEN IT reopens in late 1957, Casey's barn has been transformed. The ground floor of the main building is taken up with an enormous new box that fills almost the entire interior. Twenty feet wide and over forty feet long, ten feet high—roughly the size of two large shipping containers side by side. A house built within a house, nested like a Russian doll. The box has six doors, three along each side. Wooden steps ascend to the top level where, close under the barn's vaulted roof, a wide catwalk is suspended along the central ridge. It's an observation deck, illuminated by light coming up through six large glass panels that are cut into the ceiling of the box house underneath.

Below the catwalk are six brightly lit rat rooms. From the observation deck, everything is visible. Each of the rooms measures ten by fourteen feet and is nine feet high. Two-foot-tall partitions divide the rooms into four separate cells, each seven by five feet. An electric wire prevents rats within the enclosure from scaling the partition wall. The rooms are hermetically separated, but transit between the interior cells of each is provided by *v*-shaped ramps installed along the outer edge of the partition. The cells are labelled I, II, III, IV. There are four cells but only three ramps; cells I and IV are not connected, so the four cells effectively comprise a row, folded up. In each cell, a large artificial burrow, three feet on each side, modelled on the burrows seen in Towson, is mounted on a raised platform. Inside the burrow, five small nesting boxes. The nesting boxes are connected by internal corridors and arranged facing away from one another for privacy. In the central corner of each cell is a food hopper and water bottles. Artificial lighting simulates a diurnal cycle from 10:00 a.m. to 10:00 p.m. Temperature is a steady 65°F.

Each cell of each room is identical, except for the burrow boxes. Some burrows employ a circular design, others a linear design. Some are close to the ground, others are mounted high above the floor of the cell. At Towson, there had seemed to be a relationship between social status and proximity to resources. Dominant, successful colonies would live closer to the food supply, and would accordingly expend less effort in acquiring food. The less dominant colonies lived farther away. While he was at the Jackson Lab, Calhoun had opportunity to test this using the Black 6 mouse and constructed an enclosure comprising a long row of nesting boxes with a food supply at one end. Just as the narrow geography of Manhattan Island had first made high-rise buildings expedient, so to save on floor space Calhoun built the row of nest boxes vertically, forming a high-rise housing unit for mice. As expected, he found that mice with lower social status lived further from the food. He now applies that discovery to the four-cell enclosures. The burrows are mounted at two different heights: in cells I and II, the burrows are low (three feet); in cells III and IV, they are high (six feet). This difference introduces what Calhoun calls "an income factor," as the rats in cells III and IV will need to expend more effort to climb up to and down from their burrows. Lower burrow modules in cells I and II are expected to be preferable for this reason. Spiral ramps provide access from the floor. A supply of water and food is available in each cell at a feeding station located at the central corner of the matrix.

The aim of the experiment is a controlled analysis of the development of rat societies under different environmental conditions. The final variable is density. Where the growth of the population in Towson was allowed to take its own course, here in the barn, the demographics will be strictly controlled. Each of the four cells in a room ought to very comfortably house a colony of twelve rats—a group size Calhoun saw emerge spontaneously in both the Baltimore blocks and the Towson enclosure. One colony per cell, four cells per room, thirty-five square feet per colony. This yields an ideal population of forty-eight rats per room. Assuming that increasing the population density will accelerate the rate at which social interactions occurs, he is aiming to approximately double that number. He expects the burrows' varying heights and layouts to result in different forms of social organization, that the structure of the environment will more clearly stratify the class structure he observed in Towson. What actually happens is something else entirely.

THE ROCKVILLE study began in late January 1958. Calhoun ran his experiment in four rooms simultaneously. When the rats were first brought into the rooms, the ramps connecting the four cells within each of the four separate rooms were removed, and a pregnant female isolated in each cell. The first litters were born inside the cells in mid-February. Within each room, the pups were redistributed between the four mothers to create an even split between male and female. Each cell now contained four males and four females, plus a mother. In mid-April, after forty-five days, the mothers were removed so only the native-born rats remained, and the ramps were replaced to allow movement between the pens. As rats are sexually mature at three months, breeding began almost immediately, and after a three-week period of gestation, the second tier of offspring were born into the pens during late May. Calhoun was using the Osborne-Mendel strain of white rats, a variety available through his employers at NIMH.

Of this second generation, only sixteen males and sixteen females were allowed to remain. Each of the six rooms now contained sixty-four rats apiece. In August 1958, a third generation was born, and again, all but thirty-two pups were removed to create a target population of ninety-six rats per room. With food, water, and bedding provided, a comfortable temperature, and no predation or disease, there was relatively little mortality. As the experiment continued through into 1959, Calhoun's only intervention was to continue to remove pups to avoid excessive overcrowding, artificially maintaining the population size, which settled at around eighty rats. For the first period of the experiment, the rats established stable colonies and moved freely between the different cells. "In fact," Calhoun noted, "from watching the activity going on as I sat at the window above the room, I developed a fairly strong impression that there was some interval of time after which if a rat continued to be active it just had to get out of the pen it was then in and go elsewhere."[6] Beginning in September 1958 and into early 1959, strange behaviors began to develop.

Even with the population at almost double the optimal level, if the rats had distributed themselves evenly between the four cells of each room Calhoun estimated that a density of twenty-four rats per cell, ninety-six rats in total, should have been entirely tolerable with the one hundred and forty square feet available. But instead, the population began to cluster in the central cells.

Several factors were in play. Initially, when young rats left their birth cell to explore a neighboring cell, they were more likely to fill up the middle cells—which had two ramps, one on either side—rather than the end pens, which had only one ramp. Two entrances meant there were twice as many ways to enter the middle cells, while the end cells each had only a single entrance, making them easier to defend. In short order, a dominant male would emerge to occupy a position on the entrance ramp, converting it to an exit-only ramp. The dominant males, whom Calhoun described variously as "despots," "aristocrats," or "king-pins," would prevent rival males from entering the end pens, while collecting females behind him in his new territory.

Driven out of the high-status cells, the more submissive rats were forced to occupy the middle section, which became predominantly male as so many of the females were now trapped in the end-cell colonies. Of those two middle cells, the third had the higher burrow unit, and was therefore marginally less favored than the second cell. In time, this meant that the dominant male in the fourth cell was able to extend his territory farther into the adjacent third cell, driving even more rats into the second cell. Consequently, that second cell began to collect an ever-greater share of the total population.

Although there were food hoppers in all four cells, because the population distribution was weighted toward the center cells and the second cell was marginally more attractive, it was more likely that there would be other rats in this zone. Food was provided in the form of large pellets, held in a small hopper behind a wire grille. Gnawing pellets out from behind the grille was quite laborious, requiring the rats to spend longer than they would prefer at the feeding station and increasing the chances that two or more rats would be eating at the same time. With so many now clustered in the second cell, they began—accidentally but inevitably—to associate feeding with having other rats nearby. Calhoun hadn't planned this, but it was the type of unexpected effect he aimed for, and as the population continued to grow, it became significant. The rats were becoming conditioned to associate feeding with proximity, and this positive association became self-reinforcing.

Cell II even became the favored feeding cell for rats from other parts of the room who left their home cells to eat with the crowd, which further increased the clustering in the middle cells. Meanwhile, any subordinate males still residing in the desirable end cells that ventured into the central cells for communal feeding were likely to find their attempt to return home blocked by the dominant male. The effect was pronounced: in one of the rooms, a single dominant male occupied the fourth cell alone with a harem of twenty females. The concentration in the center grew ever-denser.

Eventually, the rats of Casey's barn became so strongly conditioned to associate feeding with company that they would only feed when other rats were present. An aggregation developed, with more and more rats clustering at the same feeding station while other stations stood vacant. Calhoun called this behavior "pathological togetherness." Within this central swarm, the rats became disoriented. As the population density in cell II increased, the intricate system of rules that usually governed their interactions could not be maintained. Social order began to unravel.

Among the early signs that something was wrong were the nests. Absorbent sawdust and wood shavings were sprinkled into the cells and burrows as substrate, and paper strips were supplied for the rats to use for nesting. During the early phase, the paper would be carried from the floor of the pen to the burrows, where it was used to construct deep bowl-shaped nests for sleeping and rearing young. But increasingly, the paper was being dropped before it reached the burrows. Inside the burrow containers, the nests were poorly maintained—frequently just a flat pad of paper instead of the structured bowl-shaped weave. Eventually, young pups were birthed directly onto the sawdust, with no attempt to build a nest at all. Few of these survived.

As with most species, mating in rats follows a formal procedure. In the wild, as in Towson, males signal their desire and fitness by performing a sort of dance on the mound of earth formed from excavated soil near the burrow entrance. The female, if in heat and receptive, will signal her receptivity by withdrawing to her burrow. The male rat will then approach the burrow, poking his head just inside the tunnel mouth but politely remaining outside. Eventually, the female emerges, and a mock chase is performed until the male overtakes the female and mounts her. She arches her back to present her genital region—the technical term for this posture is *lordosis*—and he begins pelvic thrusting, while gently holding the skin at the back of her neck in his teeth.

As the rat society in the pens began its initial stages of disintegration, one of the first things to go was this ritualized courtship. Male rats failed to properly observe the etiquette. Females who had shown no sign of being interested were pursued, often by several males at once. The tradition of politely waiting at the burrow entrance was dropped, as males impatiently followed the females into their residence. The ritualized mock chase was also abandoned, males proceeding directly to mounting and thrusting. The gentle holding at the nape became more aggressive. Females from the later generations bore dozens of cuts around their necks—evidence that excessive force was being deployed, and of excessively frequent mounting. Sexual behavior became decoupled from reproductive behavior. Males began mounting other males, and some—especially the third-generation males—became almost exclusively homosexual. These males also bore neck wounds. "In the final phase," Calhoun observed, "young rats, even recently weaned ones of both sexes, were mounted. Such abnormality may best be termed *pansexuality*."[7]

Males entering nesting sites created another problem. In the wild, nursing mothers who are disturbed by the presence of another rat will frequently transport their entire litter to another location for safety. In the early stages of the experiment, this behavior was occasionally observed when a rat wandered into the wrong nest box. But in the later phase, as nest invasion became endemic, maternal behavior became increasingly erratic, and the transporting-to-safety routine entailed its own form of imperilment:

> The mothers would take a pup out of the burrow and start toward the floor with it. Anywhere along the way or any place on the floor the mother would drop the pup. Such pups were rarely ever retrieved. They eventually died where dropped and then were eaten by other rats.

Aggression began to sharply increase. Under normal circumstances, fighting is largely symbolic, both animals rising up on their back legs and boxing at one another. The subordinate rat will retreat when it becomes clear he is bested. The more dominant victor will frequently punish the retreating rat with a nip on the hindquarters or tail, only rarely breaking the skin. In the crowded cells, the dominance hierarchy that usually stabilized social behavior was now in a state of flux. Those males who achieved a degree of social dominance lived relatively normal lives in the end pens:

> They seldom bothered either the females or the juveniles. Yet even they exhibited occasional signs of pathology, going berserk, attacking females, juveniles and the less active males, and showing a particular predilection—which rats do not normally display—for biting other animals on the tail.[8]

A rat has extraordinary bite force, and is unusual among rodents for being capable of exerting high pressure both at the back of the mouth for chewing and the front of the mouth for gnawing.[9] They are perfectly capable of damaging one another, but seldom do so. However, the crowded rats now began to deploy some of the full force of their jaws in this new behavior:

> A rat exhibiting tail biting would frequently just walk up to another and clamp down on its tail. The biting rat would not loosen its grasp until the bitten rat had pulled loose. This frequently resulted in major breaks or actually severance of the tail.

Calhoun was baffled by this, and suspected at first that the tail biting might be a displaced form of feeding behavior, rather than aggression per se. Still the violence escalated. In addition to the neck biting and tail biting, the rats would use their incisors to slice at one another's bodies in "slashing attacks": "Gashes ranging from 10 to 30mm. may be received by either sex on any portion of the body. The depth of such wounds frequently extend down into the muscles or through the abdominal wall." This level of bodily damage was unheard of. "At times," Calhoun noted, "it was impossible to enter a room without encountering fresh blood splattered about."[10]

Those remaining males who failed to establish any dominance adopted two distinct stratagems for survival. One group withdrew from interaction altogether. They achieved a sort of social sublimation, ignoring all other rats and being ignored by them in turn. Although they appeared sleek and plump, displaying none of the wounds typical for rats in the cells, their physiological well-being was achieved at immense psychological cost. As Calhoun would later explain it, "They had ceased to be rats." Alienated, asocial, and asexual, they "moved through the community like somnambulists."[11] In later experiments, these perfectly groomed outcasts would be dubbed "the Beautiful Ones."

Diametrically opposed to the Beautiful Ones were what he called the "probers." Although probers did not compete for social status, they would nonetheless engage in violence, often directed at females. When sighted by a dominant male they would always retreat. Usually young, low-status males, the probers Jack cast as juvenile delinquents. They were hypersexual and often homosexual. Packs of them would gather and pursue females regardless of whether they were in heat or of age. The harried females began to develop large, hard masses in their abdomens. Autopsies revealed these to be dilations of the uterus, usually containing a "purulent mass" and sometimes partially decomposed fetuses. The pathology report concluded that pregnancy in those afflicted would be impossible: "The general picture of the uterus was one of severe chronic suppurative endometritis, myometritis, and peritonitis with extension of the process to the fallopian tubule and ovary on one or both sides."[12]

In the cells where a dominant male controlled the territory, some breeding still occurred, but fewer and fewer young survived past weaning. By the middle of 1959, approximately six generations and eighteen months after the experiment had begun, so few young survived past weaning age that the colonies began to die out. The combination of infertility, violence, and neglect made successful reproduction all but impossible. Mortality in the middle cells reached 96 percent.

Calhoun invented a new term for this odd collection of psychopathological activities. It is a simple two-word phrase: "I have called this process the 'behavioral sink.'" He would later explain where it came from:

> Like a geomorphic sink with its decaying vegetation and stagnant water, the behavioral sink carries the connotation of social stagnation and behavioral pathology. Maximizing the individual reward arising from bodily contiguity can lead to a nearly total social collapse and disruption of individual behavior. The behavioral sink accentuates the pathology simply arising from crowding.[13]

The behavioral sink was not a single thing but a suite of different behaviors, an umbrella term to cover the striking efflorescence of pathological activities that emerged when social density tipped past tolerable limits. Just as Selye's general adaptation syndrome followed three distinct phases of stress, so the breakdown of social order in the rat rooms at Casey's barn had its stations. It began with the compulsive closeness that he called pathological togetherness; then followed maternal neglect, increased aggression, the dissolution of mating rituals, and progressed to hypersexuality, widespread intense violence, cannibalism of infants, and reproductive failure. The sink drained into extinction.

Stress experiments had always involved some form of brutalization. Selye had used toxic injections, forced exercise, and amputations. At Walter Reed, Joe Brady had encased his animals in metal chambers, administering electric shocks to create anxiety, which he then salved with various drugs. The extraordinary reactions Calhoun had witnessed were generated without invasive disturbances. Calhoun did nothing to his animals except enclose them and allow the natural density to approximately double. After that, he withdrew. What the rats did, they did to one another. Even though there was space within the rooms to allow a tolerable coexistence, in their compulsive formation of ever-denser aggregations, they seemed to orchestrate their own collapse.

TENANCY ON Casey's barn expired in 1962. For over a decade, the barns stood empty, and by the early 1970s, they had fallen into disrepair. Casey bequeathed the property to the people of Gaithersburg, and in 1977 the barns reopened as Casey Community Center. One of the two cylindrical silos was converted into an observatory. The main barn is used for dances and wedding receptions. Where Calhoun first oversaw the emergence of the behavioral sink, there are now weekly ballet classes and the Tot-Time preschool. Eugene Casey had become a philanthropist, beloved by his local community.[14] The experiments at Casey's barn would become known around the world. But for now, they were enough to impress Calhoun's new employer. In February of 1959, David Shakow, chief of NIMH's Laboratory of Psychology, recommended Calhoun for promotion to a permanent post:

> Dr. Calhoun is a person of unusual range of interest and industry. He does not fit conventional patterns of training and background and his approach to problems is equally refreshing and unconventional. [. . .] Dr. Calhoun is the kind of investigator who carries out the exploratory work in new areas rather than the careful refinement and mopping-up of an area whose terrain has already been explored. He is the kind of person who deserves strong support.[15]

CHAPTER 9

OUT OF THE SINK

UNLIKE his previous studies, which had languished in note form for many years, Calhoun wrote up his results from Casey's barn promptly. David Rioch would have been proud. In September of 1959, only a few weeks after the terminus of the first round of experiments, Calhoun announced his preliminary findings in a paper submitted to the American Psychiatric Association Symposium on Animal Behavior. "'Behavioral Sinks,' Environmental Situations which Foster the Development of a Syndrome of Behavioral and Physiological Pathology among Socially Structured Populations of Norway Rats" is a brief and soberly procedural account of the crowding phenomena observed in the barn. Although the paper makes no mention of human behavior, it is significant that he didn't choose to present to a conference of ecologists or zoologists. After all, the American Psychiatric Association wasn't especially interested in rodent behavior except as a model for human behavior.

Over the next two years at Casey's barn, he ran the experiments again and again, tweaking variables. In one version, he made food easy to access but set up the water bottles with a mechanism that required one rat to press a lever in order for another to drink. This obliges the rats to drink, rather than eat, together. Crowding developed around the water bottles as it had the food hoppers before, although the effect was less extreme. In this version, the density in the center pens was slightly lower, and the behavioral sink didn't develop with the same ferocity. Nonetheless, normal reproductive activity was significantly disrupted, and mortality rates reached eighty percent.

With more robust data now acquired, Calhoun went public in 1962. He published a technical report in an academic collection called *Roots of Behavior* among essays such as "Neurological Aspects of Insect Behavior" and "Maternal Behavior and Its Endocrine Basis in the Rabbit." But he also wrote up a more accessible account and submitted the manuscript to *Scientific American*—the same place he had read about John Q. Stewart's social physics over a decade earlier. *Scientific American* accepted his submission. Calhoun's article, entitled "Population Density and Social Pathology," was given the final features slot in the February 1962 edition, between an engineering article on friction simply titled "Wear" and Martin Gardner's regular Mathematical Games column. "Population Density and Social Pathology" is just four pages of text, but it would change Calhoun's life forever. The opening lines set the context in which he wanted his experiments to be understood:

> In the celebrated thesis of Thomas Malthus, vice and misery impose the ultimate natural limit on the growth of populations. Students of the subject have given most of their attention to misery, that is, to predation, disease and food supply as forces that operate to adjust the size of a population to its environment. But what of vice? Setting aside the moral burden of this word, what are the effects of the social behavior of a species on population growth—and of population density on social behavior?

Popularizations of science are sometimes dismissed as simply a place where scientists dumb down their work to explain it to the common reader. But popularizations were also read by experts in other fields, because every expert is an amateur in every other field except their own specialism. *Scientific American*, in particular, enjoyed a wide readership among research scientists and academics, who would learn of new developments and discoveries that might have bearing upon their own specialism. In this respect, the popularization was a fertile vector for the exchange of ideas between experts in separate fields.[1]

Magazines such as *Scientific American* not only allowed specialists to reach a broader audience, but the form permitted specialists to make broader and more speculative claims, insulated from the opprobrium of disapproving colleagues. For new fields of enquiry, where no fixed disciplinary conventions yet exist, there was no technical language with which to exclude the amateur, nor any established professional journals in which to publish. How else would Calhoun have learned of John Stewart's "social physics" if not through a popularization? And where else could Stewart have made such a bold set of suggestions?[2]

In *Scientific American*, but not in *Roots of Behavior*, Calhoun gives handles to the different types of behavioral abnormalities that develop. Only in *Scientific American* are the gangs of male rats described as "probers." Only in *Scientific American* are rats congregating around a water bottle called "social drinkers." The language facilitates comparison from his crowded rats to humans in an urban setting. That analogy with human society, and with cities in particular, overhung the work. It had been latent in his decision to first announce his findings to the American Psychiatric Association and in his decision to frame the *Scientific American* article with reference to Malthus. Nowhere but the closing lines of the "Population Density and Social Pathology" does Calhoun make any explicit mention of human society, and even then the analogy is deferred:

> It is obvious that the behavioral repertory with which the Norway rat has emerged from the trials of evolution and

> domestication must break down under the social pressures generated by population density. In time, refinement of experimental procedures and of the interpretation of these studies may advance our understanding to the point where they may contribute to the making of value judgments about analogous problems confronting the human species.[3]

Calhoun is, of course, being a little coy here. As intriguing as the results were for students of rodent ethology, few readers would have been especially concerned with the activity of crowded rats were it not for the implication that what had happened at Casey's barn carried a portent of what awaited humanity if we also continued to increase our population density. Although it remained an insinuation, the subtext was loud and clear.

Any reticence to make explicit the links to urban living certainly wasn't imposed upon him from above: the role at NIMH not only allowed him to make connections to human behavior, it required him to do so. If he retained a zoologist's caution about extrapolating directly from mice to men, it was not because humans were special but simply because any two species had their particular needs and goals. He would have been equally wary about comparing cats and dogs. Besides, there was an asymmetry in his knowledge, which he hoped NIMH might correct. He understood almost everything about the Norway rat; what he understood less was the behavior and motivations of his fellow men. Calhoun resolved to use his time at NIMH to learn about people.

THE REMODELLING of Casey's barn—which began in 1955 and would take almost three more years—left Calhoun with a great deal of thinking time, and at NIMH, Calhoun found himself among his kind of thinker. Intellectually, it was exciting precisely because institutionally NIMH was still very young.

While the importance of mental health was recognized, the priorities and methodologies required to address the issue were uncertain and fluid. Long before the ossification of roles and responsibilities and sub-subdisciplinary specialisms were established, there was, in 1955, less academic rigidity and far less administrative oversight. Researchers were hired and given roles for which no body of norms had yet been established. Among Calhoun's cohort of new employees was a young psychiatrist called Leonard Duhl, who had been charged with "long-range planning for public mental health." Duhl later remembered being nervous about accepting the role: "I didn't know what long-range planning was." It is entirely possible that no one did.

While training as a psychiatrist in the early 1950s, Duhl had volunteered for the Public Health Service and was assigned a post in northern California, just outside Berkeley, where a campaign was underway providing free chest x-rays to check for early signs of tuberculosis and other lung problems. The health authorities had been dismayed to find a low take-up, and Duhl was sent to investigate what was happening. He discovered that attendance at the x-ray clinic divided almost exactly over class lines: the healthy and wealthy got their x-rays, whereas the poorer populations—who often had a higher risk of lung disease—did not. It seemed that poverty was creating social disorganization and community disengagement, and that disengagement was worsening the health of the poor. Assisted by the local Quakers, he began to run community outfits—well-baby clinics, free legal advice, day-care programs, chicken dinner cookouts. As Duhl later remembered: "I found that if we just did a public health program by itself, we'd get little response from this population. But if we did community organization and got everybody involved, we got a lot more response." Although Duhl was obliged to pursue a neurological disease model of mental illness to secure his professional credentials, he increasingly felt that social class and living conditions accounted for a significant share of the health outcomes: "Poor people ended up in mental hospitals, while well-to-do people ended up in psychiatric therapy." He became convinced that social welfare and community engagement were essential components for well-being, that how and where people lived closely predicted their mental and physical health. He became an activist and would campaign for what he later called "Healthy Cities" for the rest of his life, working with UNICEF and the WHO, and writing speeches for Robert Kennedy on urban policy.[4]

In Len Duhl, Calhoun had found someone who could help him connect his ideas to more specific human concerns. In May 1955, Calhoun and Duhl went for lunch together in Bethesda. They talked about the importance of the environment to health, of how where you lived affected how you lived. Duhl spoke about the problems he had faced trying to solve the communication gap between health authorities and poorer communities. Calhoun, too, was becoming preoccupied with the importance of communication. He thought of signal voids in the Maine woods, and of how the rats in his studies used vocalizations in negotiating boundaries to avoid unwanted contact. He thought of how the life of the rats in the Baltimore blocks had represented in microcosm the lives of the citizens in whose yards they lived. Both men felt the problems afflicting the inner cities were complex, and inadequate channels of communication existed between the various thinkers whose input would be useful here. Disciplinary conventions obliged experts to treat separately issues that were intimately related: housing and health, city planning and psychology, culture and environment.

By the end of their lunch, Duhl and Calhoun had agreed on a plan to organize a conference of specialists to address the issue of space and environment, of how the spatial environment in which people lived impacted how they experienced their lives. In his notes that day, Calhoun tries to outline the promise and the challenge: "If we knew the properties and use and structuring space which modify mental health, effort could be made to devise proper spatial structuring and usage of space. Such action should elevate the general plane of mental health of the population."[5] Cities would be their focus. The intention was to address what they began to call "the urban condition" by combining the sort of experimental work Calhoun was doing on crowded rodents with Duhl's experience of working directly with communities. But the city was a hopelessly complex task, embedding so many overlapping functions: residential, commercial, industrial. Sociology cut across geography, economic history cut across commerce, politics cut across architecture. A city was the concrete expression of social complexity. How to approach such an issue? If disciplines were boxes, into which went the city? You had to break the problem up, build a system of specialists.

Overseeing the North American Census of Small Mammals during his time in Baltimore and Maine, Calhoun had learned the value of a distributed network of experts. But where the Census of Small Mammals had been exclusively concerned with rodent ecology, he wanted now to apply that model to the broader problem of the city. Although there were self-professed experts, such as Lewis Mumford, who claimed the city as their specialism, Calhoun felt that he, as an animal researcher, and Duhl, as a social psychiatrist, were triangulating something more complex, of which each could see but a part. It would not be anything like a formal academic department, but a loose collection of thinkers united by a shared subject matter. Addressing the complexity of the urban situation would require disciplinary diversity and methodological pluralism. There wasn't a box into which a concept like "city" would fit. Calhoun and Duhl both knew that it was improbable they were the only ones interested in the role of the environment in structuring social life, and if they had found each other, there must be more. An invisible college. So they began sending letters, setting a trapline to catch like-minded thinkers.

They drew up a wish list of people to invite: anthropologists, architects, city planners, environmental physiologists, mathematical biologists, physicists, psychiatrists, psychologists, sociologists, and zoologists. Among the invitees: Catherine Bauer, a city planner from the University of California who had drafted the Housing Act of 1937, the same bill that had spurred the Baltimore Plan; ecologist Edward Deevey, whose 1960 essay "The Hare and the Haruspex" would become a classic of population ecology; John Seeley, a researcher working on the causes of addiction; and John Stewart, whose social physics had been so important for Calhoun's version of population dynamics.

Not everyone said yes. Norbert Wiener sent a clipped reply: "While the problem of space is interesting, I find it hard to give to it the philosophical importance which you attribute to it. I am afraid I am not sufficiently moved by its importance to be able to give much advice concerning the steps you are taking." It had been Wiener who first realized that Walter Cannon's biological concept of homeostasis also applied to complex social, technological, and mechanical systems, that what we recognized as "complexity" in both living beings and socio-technological systems involved self-governing feedback mechanisms. He called that system *cybernetics*, a concept that would become an organizing principle of the nascent computer age. Wiener, a former child prodigy now trapped in a man's body, had little interest in the physicality of how information was communicated, or why real space mattered. It would take a novelist, William Gibson, to imagine what the consequences of dividing information from the world of things might look like. In 1984's *Neuromancer*, Gibson described how a virtual world made purely of information might be experienced by embodied humans living within it, and called that hybrid realm *cyberspace*.

By July 1955, Calhoun and Duhl had a short list, and a conference was scheduled for May 1956. Duhl, now appointed head of NIMH's Extramural Research Division, oversaw the organization and funding. They called themselves the "Committee on Physical and Social Environmental Variables as Determinants of Mental Health." The unwieldy title didn't stick. Duhl explained how, during their second conference in 1957, "in the middle of one of our way-out discussions, the Russians shot up a Sputnik."[6] The confluence was too cute to ignore. Informally, the group would subsequently refer to themselves as the "Space Cadets."

For the next ten years, until 1966, the Space Cadets would meet twice annually, often in a hotel near Dupont Circle, in Washington, DC, usually over two days. Chaired by Duhl, the cast of delegates was never quite the same year on year. The sociologist Erving Goffman joined in only briefly, the Scottish landscape architect Ian McHarg became an enthusiastic contributor. John Stewart was a regular, as was John Seeley, whose thinking about the causes of addiction would be influenced by his exposure to the group. At nearly every meeting, Catherine Bauer's pragmatic knowledge of city ordinances and gubernatorial politics provided a real-world counterweight to some of the flightier theoretical ideas. They were hopeful and ambitious. The 1957 minutes note that "Mr. Robert Kennedy has accepted to participate in next conference." He didn't.

While their membership altered, the numbers at each session remained quite stable: Calhoun had deliberately aimed for twelve delegates per meeting. That this was also the same number of individuals he had seen the rat colonies settle into was not a coincidence. Building his own theory of population dynamics, he had begun to formalize reasons why twelve represented the optimal group size for communication. In time, it would come to assume near-mystical importance for him. For now, the Space Cadets were Calhoun's way of integrating what he knew about ecology with what other specialists knew about the urban environment. It was also a chance for him and Duhl to impress upon their employer the value of their type of approach. At the inaugural meeting, Calhoun described their aim "to give direction to thinking of the NIMH to the broad area of environmental influences upon mental health."[7] That mission was made urgent by the growing influence of pharmacological fixes, which Calhoun explained to them as they entered their second year:

> This group has concentrated on an area which I believe is different in its focus and its approach from a whole host of other groups who are concerned with mental health, to wit, a lot of pressure to cure everything with these tranquilizing drugs. This is a big area of pressure.

> Forget about everything else and give them a goose of this stuff, and they will be happy, and you can produce it economically. I think the approach of our group is developing into something different, and I think much more meaningful.[8]

CALHOUN'S EAGERNESS to disseminate his work—through forming the Space Cadets, choosing to publish in *Scientific American*, his willingness to give media interviews—was driven by the conviction that what he had first begun to understand back in Baltimore had profound implications for human society. For a venture initially conceived as a means of killing rats in wartime Baltimore, the Rodent Ecology Project had yielded impressive returns. The finding that animals were seemingly able to regulate their population size so as to not exceed the carrying capacity of their environment was remarkable. Natural populations did not simply expand until they began to die of starvation. Instead, population size was apparently homeostatic, self-correcting. Nature had an automatic brake. Calhoun's former Rodent Ecology Project colleagues were no less captivated by this discovery, but were less eager to promote the idea outside of their field of specialty.

At various points, Calhoun, Christian, and Davis would all claim to have originated the idea that population density acted to inhibit further population growth. None of them could let it go, and it would shape their careers. When asked in 1963 who had first come up with the concept, Christian replied: "So far as I am aware, my 1950 paper was the original suggestion of density-dependent physiological control of populations. Many people have climbed on the wagon, but in the intervening years no one has pointed out an earlier suggestion of the same thing by anyone else nor have I found any."[9] Davis remembered the discovery as a group effort, emerging from the combined fieldwork of Emlen, himself, and Calhoun. He didn't mention Christian at all, and while acknowledging that Calhoun had published a note about self-regulating populations in *Science* in April of 1949,[10] Davis credited himself with the first formal statement in a paper he had published later that same year.[11]

Given all remained on good terms, how to interpret this disagreement? One way is to recognize that while all concurred that elevated population density led to population reduction, each had a different understanding of the mechanism behind, and therefore the significance of, what they had discovered. For Davis and Calhoun, it was primarily a behavioral phenomenon, with Davis emphasizing the rise in aggression and social strife and Calhoun emphasizing how the disruption of normal behaviors inhibited successful reproduction. But for Christian, the phenomenon was at root physiological, and to understand it, one must analyze what was happening in the stressed animals' bodies that led to decreased fertility and sickness. To that end, while Calhoun had been in Maine, Christian had continued his own investigations on the effects of population density. Where Calhoun looked at the social level, Christian approached the issue from the cellular level.

As early as 1952, Christian had written about spontaneous, stress-related death in an article discussing "Shock Disease in Captive Wild Mammals."[12] And he also knew of several laboratory studies, including Calhoun's, that had reduced fertility rates by increasingly population density. But Christian was skeptical of their value: "These experiments were highly artificial and the question remains whether similar effects can be observed in populations which have been permitted to grow of their own accord, either in the laboratory or the field."[13] And so, in the mid-1950s, while Calhoun had been preparing his experiments at Casey's barn, Christian had begun a search for a natural case study where he could observe what happens to an organism when "social pressure" resulting from high population density was applied. The difficulty here was that populations in the wild seldom became crowded: if density began to increase, excess animals would be driven out to adjoining land, subtracting themselves from the population.

("Emigration equals death," Calhoun was fond of saying.) If suitable adjoining land was unavailable, social conflict would act to lower the opportunities for successful reproduction, as per the Baltimore blocks or Calhoun's Towson enclosure. And hence the type of overcrowded population Christian sought was highly unlikely to occur naturally. But on James Island in 1955, he found one.

WHAT REMAINS of James Island lies in the east of Chesapeake Bay, just north of Taylors Island. When the first Europeans settled here in the 1660s, James Island covered around thirteen hundred acres. During the nineteenth century, it was home to some twenty families. There was a village with a Methodist church. But by the early twentieth century, a steady process of coastal erosion and land subsidence had reduced the area of the island significantly. The families moved away, the church burned down. Only seven residents remained by 1910. In 1916, one Clement Henry, of nearby Cambridge, decided the ever-more sparsely populated and predator-free James Island would be a good place to introduce five exotic sika deer he had acquired. The sika, a small Asian species not native to North America, flourished. Still the island shrank; each year another row of trees fell into the water. By mid-century, the small creek that once divided James from Taylors Island had widened to almost half a mile of open water. A little under three hundred acres of land remained. Sika hinds have a home range of approximately one hundred and fifty acres. For stags, that area is closer to five hundred acres. When John Christian first went out there to count them in 1955, there were almost three hundred deer on James Island, allowing each a range of less than one acre. Their numbers remained stable over the next few years. Then, one winter, the population suddenly collapsed.

About sixty percent of the island's deer died during January and February of 1958, mainly females and the young. As many carcasses as could be found were recovered. A photograph taken after the die-off shows one hundred and forty-seven skulls arrayed in a symmetrical tableau, the few antlered males arranged at either side for symmetry. Winter die-offs were not unheard of but had always been attributed to starvation—the population pinched off by exceeding the carrying capacity of their habitat. Christian demurred. "In fact, malnutrition often is an *ex post facto* diagnosis," he wrote. "The deer are dead and appear to be in poor condition—therefore they must have starved."[14] Yet when autopsied, the James Island deer were in relatively good health, with moderate levels of fat. There was food in their stomachs: wax myrtle, grasses, pine twigs, bark, and tree roots. "Their musculature was, without exception, well developed with no evidence of emaciation. Their pelage was shiny and dense, not patchy, loose, or ruffled." While not malnourished, records show the average bodyweight of the deer had declined during the years building up to the spontaneous die-off, and there were marked changes in the adrenal glands. As the island shrank, the population had continued to grow. As the population density increased, the deer became smaller and their adrenal glands bigger. Biopsies of adrenal tissue revealed signs of "over-stimulation with degeneration."

Rooted in this sort of close histological analysis, Christian's sika deer study was able to establish that the die-off was not the result of transmissible disease, parasites, or predation. And food—while in short supply—was sufficient. Instead, "physiological disturbances, induced by factors associated with high population density, probably hierarchical-behavioral, were responsible for the deterioration and death of these deer." It wasn't that the animals had outright died of crowding, but that rising levels of stress had weakened their resistance to natural shocks. A cold snap that might have been endured by healthy animals would now prove fatal. The most common direct cause of death on James Island was a gradual impairment of liver and kidney functions, leading to a failure to metabolize electrolytes. Christian described their symptoms as "physiological derangements resulting from high population density."

The population collapse was not entirely unexpected—Christian had been studying the deer "in anticipation of a die-off"—but here was confirmation in the field of the type of stress-induced population regulation he had first hypothesized over a decade earlier in Baltimore. After the 1958 die-off, the population began to steadily recover. Just two years later, offspring of the remaining deer displayed healthier bodyweights and regular-sized adrenal glands. Christian had discovered a key driver behind the population cycles Charles Elton had first described among the mass migrations of arctic lemmings back in the early 1940s. Unfortunately, Christian's chance to observe further cycles would be robbed by the waters of the bay.

Today, James Island has all but disappeared, washed into the Chesapeake. Long before their island habitat vanished, the sika spread to the mainland, where their descendants now support a thriving deer-hunting industry. In 2022, the US Army Corps of Engineers began a program to rebuild James Island with material dredged from the bay. The scheme is expected to recreate a habitat of some two thousand acres. There are no plans to reintroduce sika deer.

WHY HAD Christian's population die-off not been preceded by a descent into the sort of psychopathologies seen in Casey's barn? The layout of the four-cell enclosures Calhoun had built corralled the rats into contact with one another in a way Christian's island did not. James Island was small, and getting smaller, but each deer still had enough room to avoid developing the sort of compelled contact the pens at Casey's barn had created. Calhoun had also kept the population at a subcritical level—thinning out each new litter during the period when breeding was still rapid. He had quite deliberately not allowed severe overcrowding so as to avoid the sort of sudden die-off Christian had witnessed on James Island. By maintaining population density at roughly double the optimal level, Calhoun had kept his rats in a state of chronic social stress. The rats had responded to their confinement by developing the pathological togetherness, which in turn triggered the emergence of a behavioral sink. It appeared that severe overcrowding led to rapid but only partial population collapse, whereas low-level chronic crowding fostered psychopathologies, followed by catastrophic total population collapse. Calhoun had averted a natural die-off only to create an artificial extinction event.

Be that as it may, while Christian did not report behavioral aberrations in the deer, the postmortem character of his analytical method meant that he couldn't be sure that they had not first gone mad. Crowded populations frequently displayed behavioral derangements in concert with the physiological derangements. During the mass migrations of lemmings in Norway or mice in farmland plagues, the animals abandoned their usual timidity to break cover and emerge during daylight. Charles Elton had collected many examples of swarming rodents swimming far out to sea or running over people's feet with none of the fearful caution they normally exhibited. If the deer had been acting strangely, it likely wouldn't have been recorded. Christian was there in a strictly forensic capacity: he was not studying how they lived but why they died.

To explain how crowding caused stress, the significance of Calhoun and Webb's home range studies now became apparent. In the expanse of the Maine forests, and in the adjunct study in upstate New York, the rodents had distributed themselves to allow an even spacing of home ranges. No planning was required for this arrangement: they detected the presence of nearby animals—of their own kind or other species—and moved apart accordingly. There was a network of communication—vocalizations, scent marking—that ensured a steady state of density. Calhoun had schematically graphed this population distribution as an idealized hexagonal grid, formed, like a honeycomb or soap bubbles, as the roughly circular extent of each individual's home range was deformed by the boundary with a neighbor's adjacent home range. The actual shape would of course not be neatly hexagonal but distorted by the prevalence of foodstuffs and natural boundaries—tree stumps, rocks, streams. But for mapping purposes, the hexagonal distribution was an adequate representation. Calhoun's work from this point on would increasingly return to these hexagonal grids as a means of representing networks of communication, the crucial mechanism regulating the distribution of animals within a habitat where each individual is equally spaced from every other.

Crowding was a consequence of a population increasing in numbers within a space of fixed size—the communication network between animals within a habitat was the mechanism that ensured this situation would rarely occur naturally. For the deer on James Island, the population remained constant while the available space shrank, but the effects were the same: each animal had an ever-smaller portion of land to itself. Under normal circumstances, as population density increased the first resort was always emigration. In Calhoun's enclosures fences prevented that; on James Island, the waters of the Chesapeake. But in both cases, as the quantity of space per animal decreased, unwanted social contact became unavoidable, and stress increased. Even in the relatively commodious confines of Towson—which Calhoun had called an "artificial island"—those home ranges had abutted one another. Or rather, the dominant colonies had claimed for themselves a territory that was comfortable, and the less dominant colonies had been squeezed into the remaining space, overlapping one another and creating zones where conflict would be inevitable. Casey's barn tightened this up considerably; the home ranges now overlapped so much that every individual was constantly inside another's domain. If the home ranges were as soap bubbles, pathological togetherness represented the point at which they burst into one another, forming a single bubble, and the sense of self or ego of each individual shrank back so much as to retreat within its own body.[15]

Edith and Jack, in Calhoun's office at Northwestern University, c.1942.

Alley between Baltimore row house blocks, a reference photograph by the Rodent Ecology Project of Johns Hopkins, c.1946.

Calhoun in the Towson observation tower, c.1948. X's are photographer's marks made on the original print.

View of the central portion of Towson enclosure from the tower, c.1948. A section of the outer perimeter fence is visible in the background.

Photograph of cells II and III in Casey's Barn, c.1961. The rats are white, but their pelts are marked with ink for identification.

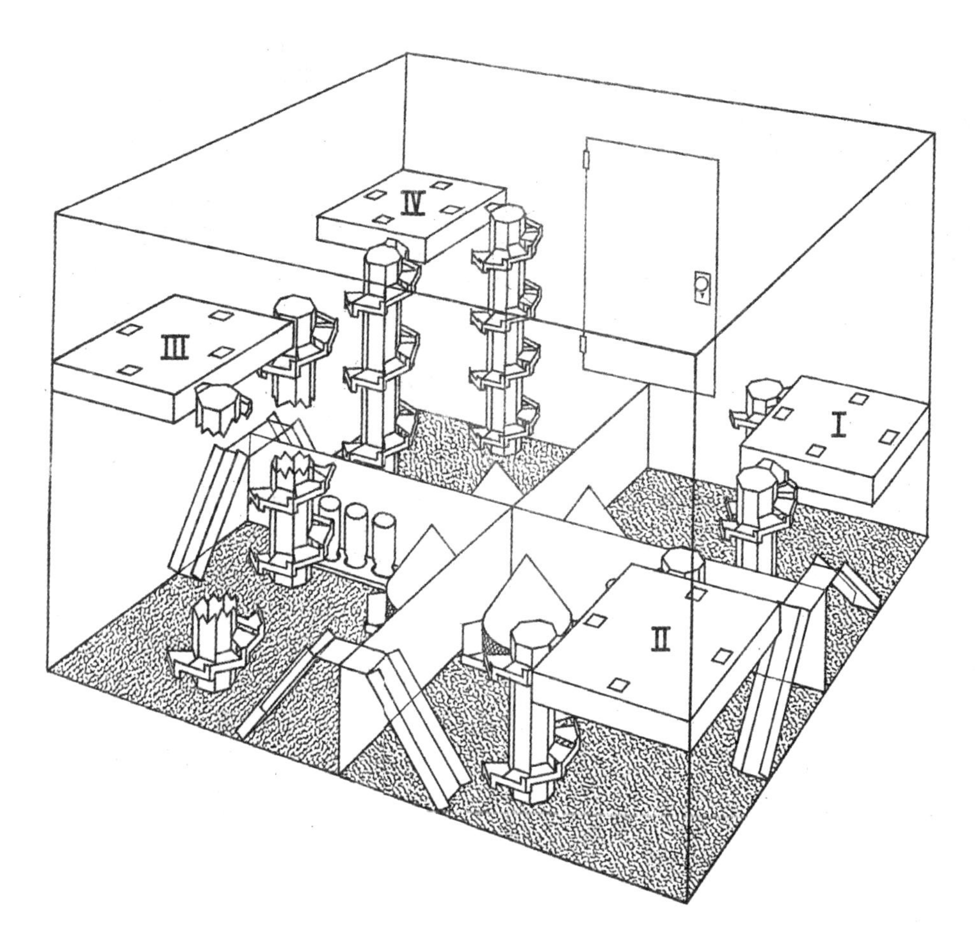

Schematic cutaway diagram showing the layout of the rooms at Casey's Barn, illustration prepared by NIMH for the 1962 Roots of Behavior collection.

Calhoun looking down over Universe 25 during the early phase of occupation, prior to explosive population growth, c.1968

Calhoun in the floor of Universe 25, c.1970. The swarms of mice following his feet are the "Pied Pipers"—females who flock towards any new object in their environment.

FOLLOWING PAGES: Calhoun inside Universe 25, c.1970. The mice at this stage are docile and pathologically drawn into dense clusters around food sources.

GINGER
ALE

4
3
2
4
5

Male mice cluster on a lab-stand ring, exhibiting infantilized need for bodily contact. These are among the "Beautiful Ones" who survived into the Death Phase of Universe 25, c.1970

Male mice lined up along a lab-stand arm in Universe 25, c.1970. Although physically touching one another, they do not interact socially.

S133,U33

HUGH BROWN
ART SERVICES

A detail showing one "cell" of Universe 33

One of the URBS team, computer technician Garrett Bagley checks an electronic portal in the velocity pen of Universe 35, c.1979.

Jack Calhoun at the Poolesville URBS Lab. On the door behind is a reproduction of the illustration used by *Innovation* magazine showing revolutionary mice, and Calhoun's "Prescription for Evolution" motif, c.1978.

PART TWO
EXODUS

CHAPTER 10

PERSONAL SPACE

WHILE Calhoun was thinking about how shrews and voles managed to distribute themselves within a habitat to avoid unwanted contact, and of what bearing the contraction of adequate home range might have on the behavioral derangements he had seen in his severely crowded rats, an anthropologist, Edward Twitchell Hall, and a psychologist, Robert Sommer, were also investigating the effects of proximity on behavior. In the late 1950s, quite independently from Calhoun and from one another, Hall and Sommer were each developing an account of how our sense of personal space played a crucial role in social activity.

When they learned of Calhoun's rats, they found their theories about "personal space bubbles" slotted into Calhoun's work, and to broader discoveries from the field of animal behavior. Ecology and sociology began to entwine. Calhoun had, of course, already begun to make tentative connections between his crowded rats and crowded humans. But it would be through the advocacy of Hall and others that his work really took off, as during the 1960s a growing number of city planners, psychiatrists, architects, and prison reformers would come to see in Calhoun's account of pathological togetherness and behavioral sinks a powerful argument for the importance of crowding as a factor in mental health, and a compelling naturalistic explanation for rising violence and civic unrest.

Over the next decade, Calhoun would find his model of the behavior of rats trapped within a city block turned back onto the behavior of people similarly trapped within an environment from which they could not easily emigrate. In mental institutions, in prisons, and in the towering concrete housing projects replacing slums in cities across America, Calhoun's rats were escaping the laboratory.

IN JANUARY 1949, President Truman had used his inaugural address to lay out an international "program for peace and freedom," comprising four key points of foreign policy.[1] The last of these involved the global dissemination of cultural and technical learning so that other, less-developed nations might benefit from America's knowledge. On account of being Truman's fourth point, this scheme became known as the Point Four Program. Twenty-five million dollars were set aside. One of the first people hired with this money was anthropologist Edward Hall.

During the 1930s, Hall had been employed by the US Indian Service, building roads and dams alongside Navajo and Hopi construction crews. When called up for the war, Hall was made an officer and attached to an African American engineer regiment. They were sent to Normandy, then the Philippines. Every job was also a chance to do his version of anthropology, closely observing the interpersonal dynamics and behavioral codes among people from different cultures and subcultures. Hall would later say, "To me people were just plain interesting."[2]

In the early 1950s, Ned Hall's role with the Point Four training program would find him advising US diplomats on how to overcome communication barriers during foreign assignments. That task involved far more than learning the language. Besides speech, there was a complex code of behavioral norms to observe and negotiate—tone of voice, posture, duration of eye contact—all of which varied from country to country, from group to group, and all of which transmitted messages. Hall called this unspoken communication "the silent language."

Suspecting that what the Point Four Program meant by "mutual understanding" was actually "American understanding," he tried to make his diplomats realize that they did not occupy a neutral position. Where previously the anthropologist's game was describing the strangeness of other cultures, Hall wanted to show his fellow Americans that they, too, were strange. Hall picked out scenarios that were amusing and recognizable. Example: If you're sat in the office with your feet on the desk, you put them down when someone walks in. If that person is a friend, the feet go back up. If that person is the boss, your feet stay down. Everyone knew this. But it wasn't written anywhere.

There's nothing surprising or counterintuitive about Hall's silent language—quite the opposite. Like observational comedy, Hall's work was popular precisely because it was so recognizable. His gift was to notice what everyone already knew but no one had thought to say, and that knack for making the apparently obvious actually obvious proved enormously popular. An editor from *Scientific American* urged Hall to write up an account of his ideas for their magazine. In 1955, Hall produced a short article on "The Anthropology of Manners." The response was so positive that he was invited to write a book: *The Silent Language* was published in 1959 to great acclaim.

Although Hall's early work had attended to multiple aspects of nonverbal communication, it would be spatial relations that increasingly preoccupied him. He was especially alert to the delicate politics of interpersonal proximity:

> We have strong feelings about touching and being crowded; in a street car, bus or elevator we draw ourselves in. Toward a person who relaxes and lets himself come into full contact with others in a crowded place we usually feel reactions that could not be printed on this page. It takes years for us to train our children not to crowd and lean on us.[3]

Being emotionally close to someone could be cashed out as a willingness to be physically close. Being physically close without any emotional closeness generated discomfort. Among the lessons Hall taught his Point Four diplomats was that personal space was culture specific, and varied geographically. As he formalized his observations into a rigorous system, he began to speak of it as a new science, and in 1962, named his theory of personal space *proxemics*.[4]

The idea of "personal space" is so commonplace now as to seem a given. But remarkably, almost nothing had been written about personal space prior to Hall. Although Hall had himself already made analogies with "what would be territoriality in lower forms of life,"[5] he had no idea that something very like his conception of proxemics had been occupying biologists for several decades. For the biologists, the idea originally emerged from the difficulty of trying to imagine how a nonhuman animal experiences the world.

In the 1930s, Jakob von Uexküll, a Baltic German from what is now Estonia, had done for the immediate exterior world of animals what Claude Bernard had for their interior world. Uexküll realized that the ongoing survival of each species depended on that organism having a particular and intimate sensory relationship with its surroundings, what he called its *umwelt*—which can be translated as "environment" or "surrounding space." When he tried to imagine what it might be like to be a completely different sort of animal—he chose a tick—Uexküll saw that whatever sensory inputs and stimuli the tick was receiving would be very different from those of a dog or a bird: the umwelt was species specific. The essay he wrote on the sensory world of the parasitic tick became a classic of philosophical enquiry, and would prove the unlikely spur for Martin Heidegger's revolutionary philosophy of phenomenology.

During the 1940s, the German-born psychologist David Katz and the Swiss zoologist Heini Hediger would similarly notice how different animals each had very distinctive spatial relationships, both with one another and with their environment. Katz called this "a network of personal space relations."[6] For Hediger, it was the "flight distance"—the invisible but definite perimeter around an animal which if crossed would cause it to startle, as if triggering an adrenal trip wire.

It would not be until the 1950s that psychologist Robert Sommer, treating schizophrenics in a Canadian mental institution, read Hediger's work on zoo animals and realized that humans, too, have a flight distance. In a 1959 article called "Studies In Personal Space," Sommer became the first to use the term in its current sense. Crucially, Sommer made a distinction between *territory*, which was fixed to a particular place, and *personal space*, which was centered on the individual and carried around with them.[7] What Hall and Sommer meant by "personal space" described a field enveloping an individual, an invisible aura. Our self extended beyond our skin, and any intrusion into this was experienced as a violation. To notice what had for so long gone unnoticed had taken a degree of estrangement. Sommer's came from working with mental patients, Hall's from working with American diplomats.

United by their shared fascination with personal space, Hall and Sommer began a lengthy correspondence, developing one another's ideas and observations. Feeling like outliers in their own disciplines, they were each amazed to discover someone else was interested in the same issue. In 1960, Hall wrote to Sommer to say how "it is encouraging and rewarding to find a colleague."[8] They were always on the lookout for what Sommer called "another one of us marginal people."[9] After Sommer tipped off Hall to Hediger's work on flight distance, Hall began reading more about territoriality in animals.

The world of animal studies opened up new ways of thinking. Zoologists had a dazzling array of species to study; anthropology had only one. Progress for the zoologists consisted in discovering the distinctive characteristics of different types of animal—calculating the home range, migratory habits, mating rituals—to build up a picture of what each species required for survival and reproductive success.

It was really only after reading the animal studies that Hall began to think of the human as an animal with fixed needs, and it became possible then to question whether, and to what extent, the habitat in which the human animal lived was suitable for those needs. When William Burt had first introduced the concept of home range two decades earlier, he had emphasized its importance for wildlife management: "How can we manage any species until we know its fundamental behavior pattern? What good is there in releasing a thousand animals in an area large enough to support but fifty?"[10] It occurred to Hall that as an anthropologist, his role was to discover the fundamental behavior patterns of humans.

While living in Washington during his stint training diplomats with the Point Four Program, Hall had become friends with a circle of psychiatrists, one of whom was David Rioch—shortly to become Calhoun's mentor at Walter Reed. That confluence was to prove significant for both Hall and Calhoun. When, in 1960, Rioch heard Hall was looking at the dynamics of personal space, he suggested Hall contact Jack Calhoun, whose studies on crowded animals might be of interest. At this point, Calhoun had not yet published anything on the experiments at Casey's barn, but he sent Hall a preprint of his *Roots of Behavior* essay and a draft of a long piece on population dynamics.

Hall was captivated. He replied to Calhoun to say he found his work "exceedingly stimulating," and the behavioral sink "downright exciting."[11] That fall, he visits Casey's barn. The experience was to have a profound effect upon him. Hall's output had always been quite lighthearted. All that would change after he first encountered Calhoun's rats.

WHERE HALL'S first book, *The Silent Language*, had been about nonverbal communication in general, his 1966 follow-up, *The Hidden Dimension*, focused specifically on how space affects our behavior. Where *The Silent Language* used anthropological case studies, *The Hidden Dimension* used a Noah's Ark of animal studies. There are vignettes about sticklebacks, walruses, crabs, and swans. There is the story of the population collapse of John Christian's James Island deer. Calhoun's work receives almost a whole chapter. Toward the start of *The Hidden Dimension*, Hall provides an extended summary of both the Towson and Casey's barn experiments, and references to "the sink" are scattered throughout the remaining chapters.

With his account of mismatched manners now plugged into the physiological theory of social stress from the ecologists, Hall sensed something terrible looming. At stake was not the occasional faux pas or social slight, but the integrity of human civilization. When the interpersonal was scaled up—when awkward interactions were summed across a population and repeated over time—it began to take a collective toll on the citizenry. At the personal level, being crowded together was unpleasant. At the societal level, it was potentially disastrous.[12]

As the site of peak social contact, it would be the city that concerned Hall the most. He increasingly came to believe that cities were deeply inimical to human welfare, a pessimism underwritten by animal studies: "Ethology and comparative proxemics should alert us to the dangers ahead as our rural populations pour into our urban centers." That phenomenon of cities acting as attractors, which John Stewart had called demographic gravitation, Hall now couched in the martial register of munitions. He wrote of the "implosion of the world population into cities," generating "destructive behavioral sinks more lethal than the hydrogen bomb." This, certainly, was a long way from an office worker with his feet on the desk.

Far happier riding his horse through the mesa near his old home in New Mexico than railcars through the metropolis, Hall felt sorry for city dwellers. As they made each crowded subway journey, were jostled through doors, endured the latent malevolence of an underground carpark or the sensory deprivation of a windowless office, the city's population was kept in a perpetual state of low-level tension, wary, alert. Taking the animal studies as his model, Hall acknowledged that tamed animals—a category in which he includes humans—can sustain remarkably high population densities assuming they are able to feel safe from one another.

In a functional sense, Hall understood architecture as a series of privacy screens, behind which one might retreat from the presence of others. A screen shut out the babble of the silent language. When there was nothing to hide behind and personal distances involuntarily overlapped, just as with Christian's deer or Calhoun's rats, overstimulation and stress ensued. Each factor was locked in an escalating process of mutual aggravation: "When stress increases, sensitivity to crowding rises—people get more and more on edge—so that more and more space is required as less and less is available."

Hall's Calhoun-colored account of the city as a behavioral sink caught the attention of the journalist and later novelist Tom Wolfe. In 1966, Wolfe spent two days in New York with Ned Hall. Here was a psychedelic-free opportunity to describe his fellow Americans in the most bestial terms. Standing beside Hall on the plaza balcony overlooking the ticket hall of Grand Central Station, Wolfe saw the rush hour commuters below through an animalistic filter: "running around, dodging, blinking their eyes, making a sound like a pen full of starlings or rats or something." In characteristically lurid style, he saw the residents of Manhattan helplessly "sliding down into the behavioral sink":

> Overcrowding gets the adrenaline going, and the adrenaline gets them hyped up. And here they are, hyped up, turning

> bilious, nephritic, queer, autistic, sadistic, barren, batty, sloppy, hot-in-the-pants, chancered-on-the-flankers, leering, puling, numb—the usual New York, in other words.[13]

Wolfe was clearly having fun, but his apparently scattergun adjectives were carefully chosen: each picked out one of the physiological and behavioral derangements Calhoun and Christian had identified from their crowding studies. Wolfe's style might be exaggerated, but his exposition of the ethology was on point.

Wolfe's piece ran that weekend in *New York Magazine*, at that time the Sunday supplement of the short-lived *World Journal Tribune*, and would later be collected as the apocalyptic final essay in *The Pump House Gang*, a collection of Wolfe's reporting on the 1960s counterculture. Through Wolfe, a new audience would be introduced to Calhoun's rats and the idea of a "behavioral sink."

In Woody Creek, Colorado, gonzo journalist Hunter S. Thompson was moved to write a brief note to Wolfe. Thompson already believed that American society was going mad, but he was excited to have a new term to describe it. "'Behavioral sink' is up in that league with my all-time, oft-used champ, the 'Atavistic Endeavor,'" Thompson enthused, misattributing the phrasing to Wolfe, and adding: "The term itself is a flat-out winner, no question about it. Every now and then I stumble on a word-jewel; they have a special dimension."[14]

While Wolfe and Thompson relished the collocation and connotations of "behavioral sink" as an addition to the literary grotesque, for Hall, Calhoun's findings deserved more serious employment. Where his earlier anthropological work had emphasized the variety and elasticity of culture norms, Hall's immersion among the animal studies would force him to confront the limits beyond which our cultural adaptability would not extend. Inevitably, the most pressing problem for personal space was how to deal with the expanding numbers of people, and when Hall spoke of the crowding, it was almost always in reference to Calhoun's experiments:

> In animal populations, the solution is simple enough and frighteningly like what we see in our urban renewal programs as well as our suburban sprawl. To increase population density in a rat population and maintain healthy specimens, put them in boxes so they can't see each other, clean their cages, and give them enough to eat. You can pile the boxes up as many stories as you wish.[15]

Just as Calhoun had calculated that the Towson enclosure could have accommodated five thousand rats if they had been housed in standard lab cages, so Hall saw the modern city as an environment built to satisfy the same criteria: the maximum number in the minimum volume.

"AS RECENTLY as five years ago, it was difficult to interest people in the significance of John Calhoun's work with rats or John Christian's studies of the consequences of animal crowding," Hall would tell an audience of the Smithsonian Institute in 1968. "Today, one can hardly pick up a newspaper without reading about a new study on the effects of crowding."[16] Certainly, much of the appeal of Calhoun's work lay in its topicality; as wide-scale slum clearances were relocating communities into high-density high-rises, and against a backdrop of rising crime and civil unrest, Calhoun's experiments were a good fit with the widespread pessimism about the future of the city. But more than that, his work fed a growing appetite for animal-human analogies. Hall's use of animal studies to explain human behavior had once been startling and innovative. But by the late 1960s, he was one of a number of popular writers exploring the animal origins of human nature. Perhaps the most influential of these was Robert Ardrey.

For many years, Ardrey had been a successful playwright and Hollywood screenwriter. That all changed when, visiting East Africa in 1955, he had been introduced to Raymond Dart, an anatomist who had identified the first fossil remains of the early hominid *Australopithecus* in 1924. Subsequent excavations of australopithecine sites turned up a disproportionate number of antelope humerus bones. Baboon skulls from the same era bore distinctive indentations. Dart made a two-pronged reading: firstly, anatomical similarity meant humanity had emerged from African apes. Secondly, these early hominids had used the antelope bones as weapons. From this forensic evidence, Ardrey became convinced both that the so-called out-of-Africa account was correct, and that our primitive ancestry had been warlike and violent. Use of weapons not only preceded but made possible the evolution of anatomically modern man. Ardrey was fascinated: here was a story more compelling than any play. He stopped writing fiction, and for the next two decades devoted all his intellectual energies into promoting what he called *African Genesis*.

Beginning on the African savannah some four hundred thousand years ago, Ardrey's interpretation of history was as a tragedy in which we were forever stuck in the third act. Conflict was all, and violence was the defining characteristic of our species. Humans were "killer apes" whose first tools were weapons: "Far from the truth lay the antique assumption that man had fathered the weapon. The weapon, instead, had fathered man."[17]

The extreme violence of the behavioral sink was exactly what Ardrey was looking for. He admired Calhoun's indifference to disciplinary constraints, sketching him as

> a maverick's maverick in the field of psychology. Physically slight, temperamental elusive in the sense that elves are hard to get hold of, Calhoun is blessed with the ability of slipping through the formidable fences of American psychology to escape without attracting undue attention.[18]

If, for Hall, the animal studies had been a way to view humans as animals with fixed needs, for Ardrey, Calhoun's work affirmed the essential bestial-ity of humanity. Ardrey's interpretation of the Casey Barn experiments was typically idiosyncratic, focusing not on the suffering of animals helplessly compelled to gather in ever-more deleterious aggregations but instead on what he saw as the reckless hedonism that drove the formation of behavioral sinks in the first place. Ardrey noted how females in the protected end pens would, when in heat, forfeit their security to travel into the center pens, joining the melee, seeking out the lower-ranking males for copulation. As Ardrey saw it, the togetherness might have been pathological, but it remained essentially voluntary. "The rats enjoyed the behavioral sink," Ardrey insisted. And so did humans: "Any consideration of space and the citizen must recognize the historic truth that people have tended to cluster in cities because they wanted to."

In Ardrey's formulation, the central instinct around which all other forms of social organization were organized was territoriality. If we understood that, everything else fell into place. Defense of territory led to aggression; sustained protection of territory led to walls, fences, enclosures, to the concentric redoubts of the motte and bailey, to walled citadels, eventually to cities. For Ardrey, the city was a concrete manifestation of what he called "the territorial impulse."

For the ecologists, of course, cities and civilization had long been antithetical to the natural world. But that harm was now turned back onto the population for which the city was supposedly intended. Ecologists made it possible to justify the bizarre claim that the city—the ultimate human habitat—was no place for people to live. Although initially safe havens, a refuge from hostile invaders, the sheer numbers now aggregating in cities was at odds with our need for a private domain. Those same territorial instincts that first created the city now imperiled those trapped within it. "We face in the urban concentration something new under the sun, something unanticipated," Ardrey warned, emphasizing how our evolutionary inheritance left us ill-suited to our present circumstances:

"We may live in our cities like ants in an ant-hill, as vertebrates we are genetically unprepared for such contingency."

Ardrey's 1961 book *African Genesis* became an international bestseller, topping *Time* magazine's list of the most notable nonfiction books of the 1960s.[19] Yet if his prose style won over audiences, it distanced many academics, who felt Ardrey's flair for dramatic clashes had led him to overemphasize the role of violence at the expense of kinship, nurture, and cooperation—all of which were also key elements of human development. Ardrey went on to write three more books expanding on his theme of innate violence and—like Hall—drawing on an increasingly wide variety of animal studies. By viewing humanity in animalistic terms, that is, *zoomorphically*, Ardrey was able to effect the sort of estrangement Hall and Sommer had achieved in their work. Unlike Hall and Sommer, however, he wasn't seeking to ease disquiet so much as provoke it.

SHORTLY AFTER Ardrey's second book, *The Territorial Imperative*, was published in 1966, English zoologist Desmond Morris published *The Naked Ape*, which was written in a matter of weeks and sold by the millions.[20] Morris agreed with Ardrey that "it is the biological nature of the beast that has molded the social structure of civilization, rather than the other way around,"[21] and added his voice to what was becoming an increasingly familiar narrative: human population growth was out of control, overpopulation and urban sprawl were damaging to the environment, cities were dreadful for nature and humanity alike. *The Naked Ape* connected Hall's personal-scale proxemics to Ardrey's grander story of savannah apes maladapted for city life, where constant crowding was tolerable only through ever-more extreme withdrawal: "By carefully avoiding staring at one another, gesturing in one another's direction, signaling in any way, or making physical bodily contact, we manage to survive in an otherwise impossibly overstimulating social situation." Unlike Ardrey and Hall, however, Morris was by then able to assume his readers would already be familiar with Calhoun's rodent crowding experiments, and mentions them almost in passing:

> We already know that if our populations go on increasing at their present terrifying rate, uncontrollable aggressiveness will become dramatically increased. This has been proved conclusively with laboratory experiments. Gross over-crowding will produce social stresses and tensions that will shatter our community organizations long before it starves us to death.

By the end of the 1960s, Ardrey and Morris had significantly contributed to the normalization of what had been a fringe position.[22] Theirs were not, of course, the era's only popular books on anthropology—this was also the decade of Frantz Fanon and Claude Lévi-Strauss, both of whom were on the same *Time* magazine list that *African Genesis* had topped—but none had the same broad and immediate appeal. The once-radical use of biologistic, and especially evolutionary, mechanisms to account for human behavior would become entirely orthodox. *African Genesis* and *The Naked Ape* foretold the rise of sociobiology, evolutionary psychology, and the entire gene-down explanatory framework used by so many public intellectuals and science popularizers ever since, including Richard Dawkins, Steven Pinker, and Yuval Noah Harari. By 1968, Ardrey had stopped scripting movies, but the opening sequence of Stanley Kubrick's *2001: A Space Odyssey* was straight out of *African Genesis*: an Australopithecus beating a rival to death using an antelope bone.

Embedded within the biologistic story was a fatalism about the possibility of change: humans were guided by their underlying animal natures. Through the lens of animal studies, it was easy and convenient to view the rising crime and civic unrest in sixties America as a biologistic reflex to rising population numbers. If such a position seemed to account for the extraordinary explosion of public disorder, it also conveniently shifted the blame from contemporary politics to evolutionary history. The wide appeal of zoological and ecological references—and Ardrey and Morris's versions in particular—was at least in part exculpatory. To the extent that violence and aggression were natural behaviors, society was absolved of responsibility for whatever it had allowed to fester in its slums.

The ecological descriptions were, in this respect, antisociological, and biology offered a refuge from politics. If violence was biological and inevitable, one needn't pursue quite so much soul-searching about inequalities and injustice; by thinking about Professor Darwin, one needn't think about Reverend King. Rather than change policies to improve equity, focus instead on tougher policing, more advanced surveillance. The racial undertones were not incidental. For those who sought it, the connotations of "killer apes" mindlessly enacting tropistic fury supplied a sufficient explanation for the riots in Harlem, Watts, and Detroit. If violence was a legacy of African genesis, then explanations rooted in systemic urban deprivation, discriminatory employment conditions, inequitable banking practices, discrepant incarceration rates, and all the unkept promises of the Reconstruction Amendments were superfluous.

CHAPTER 11

ASYLUM

AN Englishman arrived at the Saskatchewan Hospital in Weyburn, Canada, in 1951. Stepping out of the car after the long journey over endless wheat fields, he was taken aback by the sheer size of the place. Rising up three stories from the formless plain, it is the largest building between Vancouver and Winnipeg. Within its redbrick walls were psychotics, paranoiacs, chronic alcoholics, and schizophrenics. Clinical psychiatrist Humphry Osmond had travelled the almost seven thousand miles to this remote spot because he wants to give them LSD.

Osmond had been working in London when he first heard of LSD, whose hallucinogenic properties had only recently been discovered. Fascinated by the capacity to alter perception, Osmond immediately saw opportunities for psychiatry. He wanted to begin trials using LSD with his patients, but his employers were less keen. In Britain, the recently founded National Health Service had quite enough to worry about without supporting an experimental psychiatric regime. So Osmond went to Canada.

At Weyburn, Osmond began using LSD and mescaline as therapeutic treatments for alcoholism, depression, and schizophrenia.[1] The medical world was skeptical, but novelist Aldous Huxley was fascinated. In 1953, Osmond was invited down to California to supervise Huxley's first trip. In 1932's *Brave New World*, Huxley had imagined a society organized around the perfect drug. He called it "Soma"—"All the advantages of Christianity and alcohol; none of their defects."[2] The mescaline Osmond had given him wasn't quite Soma, but it came close. It would be with Huxley's help that Osmond later devised a name for this new class of drugs: psychedelics.

At Weyburn, inspired by the perceptual distortions induced by psychedelics, Osmond became increasingly focused on the way physical space affected the well-being of his patients. Alongside the therapeutic value of LSD for his patients, Osmond also thought hallucinogens could help their carers gain insight into the perceptual experience of the mentally ill and regularly encouraged his staff to try LSD. The asylum became a research laboratory. While Calhoun was building enclosures to investigate how physical surroundings affected the behavior of rats, Osmond found in the asylum a similarly closed environment, a place where it was possible to study how space affected behavior. Although as yet unknown to one another, their work began to converge.

ON FIRST arriving at Weyburn, Osmond had been struck by the imposing physicality of the Saskatchewan Hospital. But once he got inside, he was appalled. The interior was a disorienting clash of long narrow corridors and cavernous atria. In the middle of the vast Canadian prairies, the patients were perversely compacted into a space entirely unsuitable for their needs. In the private areas they felt confined and isolated, in the social areas they felt exposed and vulnerable. "It would be heartless to house legless men in a building which could only be entered by ladders," Osmond argued, but "so little thought has been given to the care of the mentally ill that buildings far more detrimental have been foisted on them."[3]

It hadn't always been this way. In letters to Huxley, Osmond contrasted the relatively modern Saskatchewan Hospital with the asylums of the nineteenth century: "Not simply relatively but absolutely our standards of building, furnishing and staffing are still below those of the 1850s. It is an appalling and almost incredible thought, but it is true."[4] The structures that replaced them were "monstrous and unsatisfactory" and "testimony to the failure in communication which has existed between architect and psychiatrist."[5]

He began to lay out principles any good design should meet, such as "Patients Must Not Be Overcrowded," "The Provision of a Path of Retreat," "Need for a Private Place," and "The Preservation and Limitation of Choice." He began petitioning the provincial government for funds to remodel the layout and hired an architect, MIT-graduate Kiyoshi Izumi.

When Izumi came to visit the site, Osmond suggested taking LSD would be a good tool to help him get closer to the mindset of the patients—to perceive the space as a schizophrenic or an addict undergoing withdrawal might experience it. This certainly wasn't the way they did architecture at MIT, but Izumi consented. Architect and psychiatrist duly dosed up, they set off to explore the wards.

Izumi described how his senses were heightened, time and space distorted. The institutional convenience of wipe-clean gloss paint of uniform color on almost every surface now disoriented him; floor, wall, and ceiling blurred together, became insubstantial. He found the glare from polished floors and reflective tiles intimidating, and the "hard" corridors that funneled everyone toward him positively frightening: "I felt that the corridors should be 'soft,' 'absorbent,' and even 'resilient,' so it could bulge out where necessary to allow another person to pass."[6] A transom over a door became the suspended blade of a guillotine. As useful as these insights were, Osmond's team needed more rigorous perspectives on the relationship between the physical environment and mental health on which to base their new design.

As word spread of Weyburn's radical program, it began to attract adventurous researchers. Among them, Robert Sommer. It was here at Weyburn that Sommer developed his theory of "personal space." He began to notice how the layout of furniture, the lighting, the placement of walls, doors, windows, all affected how groups interacted. Osmond suggested taking LSD might be a good tool for thinking about the issue, but Sommer thought he would try some observational studies first. He watched how the patients used the rooms, whether they clustered at the center or cleaved to the walls. He designed comparative experiments by altering the configurations of six chairs around a table, monitoring how some arrangements drove people apart while others seemed to induce gatherings. Sommer called his work "the ecology of privacy."[7]

AT WEYBURN, the institution became a laboratory for studying how institutions worked. Despite the fact that, as Osmond put it, "mentally ill people have been studied far more closely and intensively and in a greater variety of ways than almost any other group of people,"[8] they were amazed to discover there was almost nothing written on institutional design to improve care of the mentally ill. As Sommer later reflected: "More was known about the design of zoo cages and chicken coops than about the design of hospital wards."[9] Zoo animals were an expensive asset, chickens were a valuable commodity. People liked zoo animals, and people liked chicken. Far fewer seemed to like mental patients, and consequently very little effort had been made to improve their habitation. And so it was to the zoo that they turned.

For Osmond and Sommer, the most valuable insights had been provided by Heini Hediger, the Swiss biologist and director of Zurich Zoo who had developed the notion of "flight distance" Ned Hall would later plug into his science of proxemics. Responsible for ensuring the animals under his stewardship were fit to exhibit, Hediger found himself with a dilemma:

maximizing comfort for the animal meant designing an enclosure in which the animal was scarcely ever seen, while exposing the animal in a bare cage was likely to be extremely distressing. Visibility was purchased at the expense of welfare. A zoo presented a version of the same problem Calhoun and John Paul Scott had encountered: How to make the animal observable while preserving as much behavioral expression as possible?

For Hediger, in order to provide the best possible artificial environment for an animal it was essential to understand the natural behavior of that species. Failure to orient the artificial environment toward fulfilling these basic biological needs led to pathologies. He documented the emotional deterioration of animals improperly housed that could be seen in their repetitive movements—the nodding of a bear's head, a tiger pacing against the front of the cage. Animals noted for intelligence and activity suffered most acutely from a lack of stimulation and physical restriction, culminating in listlessness, depression, and outbursts of aggression. When Osmond read this, he recognized many of the same behaviors in his patients.

What Hediger called "zoo biology" led to some surprising conclusions: while it was of course essential to provide an animal with the correct food and companions, when it came to physical space, increased quantity was seldom the most important factor. Where previous reformers had sought to create "barless zoos," open enclosures weren't always the best solution. Animals needed a space—a territory—to call their own, and while a cage might suggest imprisonment to the human, this symbolism was meaningless to an animal seeking security and privacy.

Combining Hediger's account of zoo animals with Osmond and Izumi's drug-fuelled simulation of psychosis, the Weyburn team set about producing guidelines for the design of institutional spaces. Osmond called the new approach "socio-architecture."[10] One principle was that the quality of physical space was more important than its quantity. This might mean a shift toward providing smaller rooms, even as little as fifty square feet of floor space, in place of the cavernous dormitories and wide corridors, whose scale felt overwhelming. Large spaces also made it impossible to avoid unwanted social contact. The dorms of the Saskatchewan Hospital were a barless zoo.

Following Sommer's work studying personal space and social dynamics, Osmond suggested that a space could be classified in terms of how it distributed its occupants. The function of a hotel lobby or railway station was to move people around, to avoid too much contact and maintain distance. These were places you moved through in order to get somewhere else. He called such spaces *sociofugal.* A *sociopetal* space did the opposite, encouraging interaction and building a group identity. Sociopetal spaces, such as a public bar or family home, drew people in, nurtured a sense of belonging.[11] Hospitals such as Weyburn, with their one-hundred-and-fifty-bed dormitories and communal dayrooms linked by receding corridors, were sociofugal. Living there was like living in the ticket hall of Grand Central Station.

The remodelling of Weyburn would be modest, but significant. Following Hediger's suggestions, they divided the larger spaces using carefully placed screens, reducing the sense of overcrowding by providing places of social retreat while still permitting voluntary interactions. The solution was inexpensive and easily replicated: "What I like about it," Osmond wrote to Huxley, "is that it derives from the application of simple psychobiological principles which can be easily understood."[12]

Meanwhile, Izumi was given the opportunity to deploy their findings in an entirely new building: a psychiatric facility was planned in nearby Yorkton, a hundred and fifty miles northeast of Weyburn. His initial design proposed a circular form, consistent with the sociopetal concept. Corridors were avoided. There were small retreats around the periphery, a communal zone for large group activity in the center, and, in the intermediate zones, semiprivate spaces for small groups to interact. The plan was rejected, ironically because government grants were awarded based on compliance with existing building codes for hospital layouts. Izumi complained they had found his design "too far out."

A compromised plan proved acceptable, and the Yorkton Psychiatric Center opened in 1963. Yorkton was everything Weyburn was not—warm, cheerful, domestic, allowing for privacy and social interaction on the patient's own terms. It comprised of several small rectangular buildings that provided patients with as much privacy as possible, each with their own room. Social spaces were organized according to a modular design, with seating provided for a small number of individuals to create "spatial relationships that reduce the frequency and intensity of undesirable confrontations." Consistent with Osmond's readings in animal psychology, patients were distributed into groups of no more than ten to try to minimize any feeling of overconcentration or overcrowding.

The project to improve conditions at Weyburn proved influential. The Saskatchewan Hospital and Izumi's Yorkton plan became exemplars for institutional design. Osmond and Izumi duly became involved in a collaboration between the American Psychiatric Association and the American Institute of Architects, investigating the value of hospital design as an additional therapy in concert with medication and traditional psychiatric support. From this, a compendium was published, *Psychiatric Architecture*, a book in which the Saskatchewan plan had a prominent place—Osmond providing two of the papers. In 1959, the project's director declared that their work had nurtured "the development of effective communication between the two main professions concerned—psychiatry and architecture."[13]

Osmond's mission to convince his fellow specialists of the importance of physical space to mental health had been a success. His mission to convince them of the value of psychedelics, less so. In 1966, use of LSD in scientific research was banned. By 1968, LSD had been declared a Schedule I substance, partly because the drug was deemed to have no accepted medical use. This was especially hurtful to Osmond: always a radical, the legislation now made him an outlaw. The therapeutic use of hallucinogenic drugs was discontinued by fiat. "It is said that crime has risen greatly in recent years," he complained, "and one wonders whether this is a propitious time to add a whole new series of crimes to the burden of an already overladen police and magistracy."[14]

Yet even as they were rejecting mind-altering therapies, the medical establishment was becoming increasingly reliant on an ever-expanding pharmacy of sedatives and anti-anxiety medications. "Medicine," Osmond said, "is in tune with the morality of the age in which it is practiced," and the moral indignation toward LSD was in no small part driven by a fear that its use might alter the very social values the law sought to preserve: "The psychological changes resulting from drug use are those older folk frown upon and sometimes find repugnant and frightening, in contrast with such acceptable social tranquilizers as alcohol or barbiturates." Mere sedation wasn't going to change the world, and that suited the Establishment just fine. Though disappointed, Osmond wasn't surprised. As he saw it, the official position on LSD was "that of all establishments everywhere when faced with innovation. It consists in saying, 'No, you don't.'"

WHILE SOMMER had been formulating his theories of personal space in Weyburn, in New York State, a Dutch Indonesian psychiatrist called Aristide Esser had also been thinking about the role of the spatial environment in the treatment of the mentally ill. Like Osmond and Izumi, he found himself relying on animal studies to help him understand the social dynamics of his patients. Working at the Rockland State Hospital in Orangeburg, New York, Esser had noticed that although the wards were generally peaceful, there would be occasional eruptions of violence. When he read about "pecking order" and "dominance hierarchies" in caged hens, he wondered if the aggression he was seeing might be understood in similar terms.

There was no shortage of opportunity for observation: patients at Rockland spent about twelve hours a day contained within the open ward, for an average stay of six years, during which time the resident population remained fairly stable. Elsewhere in the hospital, there were dedicated laboratories where the patients could be examined and tested. Esser realized that for his purposes, the closed environment of the psychiatric ward was already a laboratory. From a booth in the middle of the ward, concealed behind one-way glass, he set about monitoring the movements and activity of his patients every bit as meticulously as Calhoun had watched the rats at Towson.

At regular intervals, the position and posture of every patient was recorded and any interactions noted, coded, and logged on a map of the ward divided into three-by-three-foot grid. Just as in the animal studies, a dominance hierarchy emerged that was expressed by the use of space. Although most patients withdrew to a particular spot and established a defined territory—a corner, say, or a table—which they stayed within and aggressively defended, he noticed that the more confident patients roamed freely around the ward, using the entire floor. The most timid patients were also mobile, but for quite different reasons. If they took a chair, they were likely to be routed—forced, as Esser put it, to "continually cede their place to higher ranking individuals, and thus are chased around all of the available space."[15] The squabbling and fighting would chiefly occur when these roles were challenged or contested, but as the dominance hierarchy was largely settled, the outbursts of violence were sporadic and infrequent.

Esser's result was a working definition of territoriality on a psychiatric ward that drew directly from the study of animals. Crucially, he showed that the sporadic violence wasn't solely a consequence of their psychological disorders but had much to do with the space in which they were housed. Far from being random, aggression was functional. It established rank within the ward, and rank conferred certain privileges, such as preferential territory or freedom of movement. Along with the physical walls that divided the ward, the dominance gradient established an invisible second set of barriers, limiting the space available to each patient according to their social rank. In an institution, where there was little floor space and fewer personal possessions, fighting served to protect an individual's space against intrusion and preserve the stability of the dominance hierarchy.

In the outside world, the role of territoriality was harder to separate out, and likely played a less important role in the lives of people with outlets other than violence. But in the closed environment of the hospital ward, space was everything:

> Once the normal human range is restricted, as happens with people who live in an enforced community, such aspects of group interaction as aggressiveness can be related to territorial behavior and the fight for a stable position in the dominance hierarchy.

Eruptions of violence became more likely when density increased, or when individuals felt unable to retreat to a protected and safe space. Social strife would lead to competition, competition would lead to aggression. Esser suggested maintaining low densities and ensuring less confident patients were allotted their own nooks.

Esser had been able to disclose these patterns because of the particular circumstances he had to work with: a closed environment, unimpeded observation, and, perhaps most significantly, the limited behavioral repertoire of his subjects. Deprived of "his normal means of maintaining his social position," Esser observed, the "chronic mental patient only has his physical presence with which to attain his goals."[16] Although he was studying very disturbed individuals within a highly artificial environment, Esser felt the mechanisms he had uncovered would have bearing on the behavior of individuals in any setting where they were unable to avoid one another. A prison, a school, a city. In fact, it was precisely the regimented and simplified arrangement of life for the patients that made them such good subjects for studying the underlying mechanics of human social behavior. What was obscured by the complexity of the city was laid bare in the simple environment of the ward, which offered a unique opportunity for understanding this complex process of social ordering in relation to space.

Largely incapable of dissembling or concealing their feelings, the chronically mentally ill became precision instruments—"sensitive probes"—for detecting the role played by dominance and territoriality. The observations at Rockland were valuable precisely because they were of institutionalized humans. Up in Weyburn, Osmond and Sommer had come to a very similar conclusion. As Osmond explained, the sociality of the mentally ill was central to their predicament:

> These sick people have only one quality common to all of them without exception: *rupture in interpersonal relationships resulting in alienation from the community, culminating in expulsion or flight.* They are to a greater or lesser extent socially isolated. The psychiatric ward then has to be designed to care for people whose capacity to relate to others has been gravely impaired.[17]

That inability to relate to others was paired with an inability to articulate their own needs. Many of those worst affected by the institutions in which they were housed had no option to leave, and a very limited capacity for direct communication. Izumi had stressed how this predicament made helping the institutionalized patient especially urgent:

> Through sheer inability of a kind, he is "forced" into situations, through circumstances beyond his control, whether he is rich or poor, strong or weak, learned or untutored. Then, unlike the "normal" person, his ability to cope with the situation may be limited.[18]

In a sense, the mental patients exhibited grossly amplified versions of the same problems afflicting the general population, and whatever could be done to help those living within the institution ought to apply outside their walls.

Although Sommer and Osmond in Weyburn and Esser in Rockland were thinking first of institutional reform, they saw in their patients a chance to gain a more general understanding of the principles that governed human reactions to the physical environment.

Basing their work on the study of zoo animals, Osmond and Sommer had shown how the design of a closed environment might reduce anxiety and tension. Esser had taken his cue from pecking order in hens, and territoriality, finding that aggression and violence within a closed space was—at least to some extent—related to the distribution of persons and barriers within it. If you reconfigured the space, you might expect to reduce or increase the chance of conflict. In both cases, they had found that what held true for Calhoun's rats held true for humans deprived of the opportunity to migrate and largely incapable of articulating their distress.

Sommer made the comparison explicitly, insisting that "spatial as well as social orders crumble under the onslaught of crowding. The result is extreme social disorganization of the type described in rats and mice by Calhoun."[19] There was nothing dehumanizing about comparing mental patients to animals, or perhaps it was truer to say that dehumanization was not obviously pejorative. By thinking of their patients in animalistic terms, Esser at Rockland and Osmond and Sommer in Weyburn were generating more sympathetic and nuanced insights into the care of their patients than were being achieved by accepted orthodox approaches.

IN 1968, Sommer was invited to a meeting of the American Association for the Advancement of Science in Dallas organized by Esser. At Dallas, Sommer found a who's who of researchers working on space and behavior.

David Davis was there, talking of the physiological effects of crowding; Ned Hall, on the culturally varying definitions of "trespassing." Robert Ardrey was a discussant.

Center stage was Jack Calhoun, whose lengthy paper on "Space and the Strategy of Life" was chosen to close the conference. In his introductory remarks, Esser emphasized how crucial Calhoun's work had been in encouraging the integration of human and animal sciences, bringing together ecology, ethology, and psychology, and proclaimed: "We are witnessing the emergence of a social biology."[20]

Insights generated by ecology and ethology were invigorating the study of human behavior. When Hall, an anthropologist among ecologists, was asked by the audience why he used animal studies so much, he replied:

> I believe that the more we learn about animals the more we will know about man. If there was ever a forgotten fact, it is that man is first, last, and always an animal, a biological organism. . . . I learn more about man here, than by going to most anthropological meetings. My message to this audience is: Please keep the data coming.[21]

CHAPTER 12

PENITENTIARY

PRISONER numbers were at almost double the intended capacity when, at around 9:00 a.m. on Thursday, September 9, 1971, the rioting began. Inmates had been complaining about overcrowding and poor treatment for months, and just the previous day, a correctional officer was injured while breaking up a fight. A humid Indian summer had peaked at eighty-six degrees earlier that week. Attica was gravid with violence.

One thousand two hundred prisoners joined the revolt, over half the population. Forty-two staff were taken hostage. On the fourth day, the National Guard was called in, helicopters buzzed the yard, sharpshooters picked off prisoners from the catwalks. Before the uprising was quelled, one hundred and twenty-eight men were wounded, thirty-nine fatally, including ten of the hostages the National Guard had been sent in to save. The chaos and brutality of the Attica Prison Uprising shocked the American public. But in Washington, DC, lawyer and activist Ron Goldfarb was not surprised at all. For several years now, he had been expecting something like this.

RON GOLDFARB was collecting evidence for a class-action suit on behalf of pretrial detainees being held at the District of Columbia Jail, a brooding granite structure just two miles from the Capitol Building. The current jail was within a small parcel of federal land where the indigent, the insane, and the infected had been held ever since the nation's founding. A poorhouse once stood here, then a workhouse, then a madhouse, and by 1872, this prison house. Once containing only the most dangerous criminals, the population of the DC Jail was now comprised of minor offenders and the recently convicted awaiting transfer. The vast majority were there because they could not afford to post bail and were being held until trial, a process which often dragged on for many months. Conditions were so awful that pretrial detainees frequently pled guilty to their charges simply to get out of the place.[1] Originally designed to hold fewer than seven hundred inmates, by 1970, the jail held more than twelve hundred men, often "doubled-decked" in cells that would have been cramped even for solitary occupancy. At just six by eight feet, the floor area of each cell was, as Goldfarb was fond of saying, the same size as the surface of a pool table.

By the time Goldfarb brought his case, opinion was near unanimous that the American prison system needed to change, although there was less agreement on how. Several months before the Attica Uprising, in January 1971, *Time* magazine ran a cover story on "The Shame of the Prisons," reporting that "a growing number of citizens view prisons as a new symbol of unreason, another sign that too much in America has gone wrong." Even President Richard Nixon, who had been elected on a law-and-order ticket, had complained that "No institution within our society has a record which presents such a conclusive case of failure as does our prison system."[2]

Among those calling for reform, Goldfarb was hardly the first to use the constitutional prohibition against "cruel and unusual punishment," but he reckoned he had found a fresh angle on the Eighth Amendment. Compassion for the incarcerated was slight, but Goldfarb was not planning on appealing to compassion. He was going to appeal to biology. For several years, he had been looking for scientific studies that might help him make the case that the discomfort experienced by prisoners in an overcrowded environment constituted a substantive harm. When, in the mid-1960s, he discovered Calhoun's work, he realized that stress-induced derangements were also injuries, cruel punishments that ought to be covered by the Eighth Amendment.

"There is no room to pace," Goldfarb wrote of the DC Jail in a *New Republic* article from 1966: "The men lie and sit around . . . fester, corrupt each other."[3] Goldfarb made clear the link between crowded pens and crowded penitentiaries:

> When animal societies reach a critical territorial or minimum space density they cannot survive. If animals have this recordable physical reaction to crowding, consider the corresponding physical and psychological reactions in comparable circumstances of the human animal.

Calhoun's crowding studies seemed to offer impartial and impassive evidence that at least some of the violence seen within prisons was being generated by the conditions in which the prisoners were kept. One needn't admire prisoners or even feel sorry for them, but if violence was a biological response to crowding, neither could their behavior behind bars be attributed entirely to bad character. As currently configured, prisons were generative of the sort of antisocial behavior they were intended to deter.

In January 1971, Goldfarb approached Calhoun directly: Would he consider acting as an expert witness for the case? Goldfarb offered to arrange a tour of the DC Jail so that Calhoun might see firsthand the sort of conditions the prisoners were living in. Calhoun agreed, and a jail visit was scheduled for October. Alongside Calhoun, Goldfarb assembled a group of experts. Among them was Karl Menninger, the eminent psychiatrist and author of *The Crime of Punishment*, a book in which he had urged that criminality be considered as a psychiatric disorder and treated accordingly. Also touring the jail that day was Robert Ardrey, who had recently published *The Social Contract*, which took in Calhoun's work on rats and Esser's treatment of space in psychiatric wards and linked both to how long-term prisoners understood human proximity as a threat, always preparing for ambush: "The personal space of the violent inmates was *skewed to the rear*."[4]

A photographer and reporter from *The Washington Post* turned up, partly because Goldfarb got lucky with his timing: the Attica Uprising had taken place just four weeks earlier, and—for now—prison reform was big news. "Almost everyone seems to agree that our prisons are terrible," the *Post* wrote: "Human prisoners in the United States are more carelessly handled than animals in our zoos, which have more space and get more 'humane' care."[5] Outside the jail, Goldfarb made it clear to the *Post* reporter that today's visit was an opportunity to make solid connections between Calhoun's rodent studies and human society: "For the first time," he announced, "these extremely important experiments on the behavior of mice will be adapted to the behavior of man in an overcrowded jail situation."[6] For his part, Calhoun avoided speaking to the reporters. Nor did he conduct his own interviews with the prisoners. He was by now well accustomed to mute interrogation, and just as John B. Watson had circumvented the self-reporting of his subjects to observe only their behavior, so Calhoun's ecological methods did not require detailed individual histories of the inmates. To observe how they used their environment was enough. He spent only three hours inside the walls that day but would think about the visit for many years to come.

Socially, the DC Jail offered a stark dichotomy: total isolation or total immersion. Prisoners were allowed out for exercise and meals for just two hours each day. As so many were double-decked two per cell, Calhoun reckoned the floor space per person at just twenty-eight square feet each. Calhoun noted that dogs used for medical research were allotted a minimum of thirty-six square feet. As with the mental hospitals, humans had less protections than animals with regards to their spatial requirements.

If an inmate had their own cell, they were alone for twenty-two hours of each day. If they shared a cell, they must endure continuous engagement with only one other person. To describe the double-decked, he reached for Sommer's language: "Privacy is totally destroyed since each inmate's minimum personal space is continuously encroached by the other one."[7] Many of those facing only minor charges were housed in communal dormitories. Against the understimulation of the cells, inmates in the dorms faced an even more destructive stimulus overload: constant shouting, blaring radio, a television playing continuously, the churning proximity of stressed and bored strangers. Neither the cells nor the dorms allowed for controlled social interaction: forced into solitude, forced into a dyad, or forced into a crowd.

A seasoned field ecologist who once traced filigree rat runs through the snowy yards of Baltimore row houses, Calhoun now directed that same scrupulous attention to the DC Jail exercise yard: "The fact that only near the door from the jail is the grass worn thin or bare, while over most of the yard the grass is several inches high, indicates that very ineffective usage of this space is made." The inmates had space, but barely used it. Why? The jail environment seemed to be clipping the expression of normal behaviors, much as he had seen in his rat studies. As Calhoun put it, the prisoners' "scope of existence" had narrowed, limiting their opportunity for action. In this simplified environment, they were displaying simplified reactions: a grossly attenuated behavioral amplitude that oscillated between sullen retreat and violent release.

While locked away, such men could be ignored—and were. But these were not lifers, and thinking ahead to their inevitable release, Calhoun stressed how it was in everyone's interest that they did not leave the institution more socially misaligned than when they came in. From what he knew of how his rats and mice reacted to extraction from a crowded pen, such trauma was not easily divested. He predicted that when returned to freedom, these men would exhibit either "extreme withdrawal on the one hand or explosive inappropriate reaction to society upon release into a more complex world."[8] For Ardrey, the jail seemed purpose-built to foster violent behavior: "Virtually every step possible to promote the aggressive potential in prisoners has been taken with the utmost diligence."[9] Menninger, who was already opposed to prisons, agreed with Goldfarb that the social conditions in the jail violated the prisoners' Eighth Amendment rights: "To deprive a man of decent social relationships, palatable food, normal sexual relations, friendships and constructive communication does not strike the law or lawyers or even the public conscience as cruel or unusual."[10]

As Goldfarb's experts compiled their notes from the visit, his case for prison reform was beset by numerous delays and would not be heard until 1975. In the meantime, almost exactly a year after the day of the guided tour, the eruption of violence he had been warning about finally occurred. Early in the morning on October 11, 1972, Goldfarb received a phone call. At the DC Jail, two inmates had taken a guard hostage. A few hours later, it was fifty inmates and twelve hostages. As unrest spread, the prisoners chanted, "Attica! Attica!"[11] Goldfarb was called in to negotiate. Remarkably, he was able to persuade District Court Judge William Bryant to meet with the rioters, and six inmates were nominated to attend an emergency late-night court session. Before the judge, they compared their treatment to that of caged animals. One Robert N. Jones, awaiting sentence, said: "I feel I'm being treated like an animal and I don't belong here." Terry Burgin, scheduled for parole: "We don't want to be locked up in no cage." "They don't think we are people," said Alvin McCoy, serving five to fifteen for manslaughter, "like we lost all sense of being a human being."[12] Judge Bryant listened attentively, offering amnesty and legal assistance if the uprising was wound down without further violence. Later that same night, the standoff ended peacefully.

It would be the same Judge Bryant who later heard Goldfarb's case against overcrowding in 1976. Bryant was persuaded, ruled in Goldfarb's favor, and demanded improvements be made to the DC Jail within fifteen days. Noting Calhoun's testimony that laboratory dogs were allotted thirty-six square feet each, he specified that each inmate receive a minimum of forty-eight square feet of living space. The *Washington Post* described the measures as being "remarkable mostly for their being unexceptional: they would impose no more than a reasonably decent standard of living for human beings."[13] For all that, when Bryant made an unannounced inspection later in the year, he discovered few of his orders for easing overcrowding had been followed. Seemingly, the district authorities were stalling, waiting on the completion of a brand-new facility, which, they insisted, would deal with the current problems. The early 1970s were the beginning of a vertiginously steep rise in American incarceration, and Goldfarb glumly predicted that the new DC Jail would be just as overcrowded as the old one. He was proved correct almost immediately: the new DC Jail opened in 1976 with capacity to house nine hundred and sixty prisoners. The average daily population that year was well over a thousand.

THROUGH GOLDFARB'S use of Calhoun's work, overcrowding had become a lever by which to force change in the prison system. Unlike more numinous harms such as trauma or discomfort, overcrowding was measurable and it operationalized suffering, while well-established housing standards across the United States meant it was easy to calculate objectively as a simple ratio of the number of individuals per unit area. As Calhoun saw things, if those who wished to cage animals were bound to abide by agreed-upon minimum space requirements, those who caged humans ought to be held to the same. With a nod to the welfare restrictions animal researchers worked within, he pointed out that the DC Jail already violated the existing DC Housing Code.[14] But while Goldfarb was glad the ecological arguments for change had landed, he knew that relying on overcrowding was a gamble. The problem with the crowding problem was that there was more than one way to solve it.

Reformers such as Menninger and Goldfarb were aiming for systemic overhaul. Menninger's position that crime was a psychiatric problem meant focusing on rehabilitation and therapeutic care, leading, ultimately, to the abolition of prisons altogether. Goldfarb's position was less radical, but he wanted to do more than simply modify the floor plans. His aim was a "non-architecture" approach that would see expanded programs of work release, halfway houses, bail reform, addiction treatment centers, and juvenile homes, all of which, he argued, could "lessen the demands for old or new prisons."[15] Both agreed that the prison population was far too large, and so their call to reduce overcrowding was really a call to reduce the prison population. But there was another way to reduce overcrowding.

Accepting the Republican nomination for the presidency in 1968, Richard Nixon had described the nation as "plagued by unprecedented lawlessness . . . If we look at America, we see cities enveloped in smoke and flames. We hear sirens in the night." Street crime had reached such a pitch that citizens feared to leave their houses, avoiding public transportation and city parks. The solution? More punishment. Too many criminals were getting away with it, or receiving inconsequential and lenient sentences, and hence his claim that "No institution within our society has a record which presents such a conclusive case of failure as does our prison system."[16] It should be clear that when Nixon spoke of the "failure" of the prison system, he had in mind something very different than Goldfarb. To reformers such as Goldfarb or Menninger, the overcrowded prison was evidence that too many people were being incarcerated. But to law-and-order crusaders such as Nixon, it was evidence that we weren't building enough prisons.

Goldfarb's victory proved pyrrhic: his wish to reduce overcrowding was answered, but not as he might have hoped. Just as Goldfarb had feared, by focusing on what was wrong with the physical environment, he had allowed the authorities to reply with a physical fix. Overcrowding was seen as a logistical rather than a systemic problem, and during the following decades, a vast increase in incarceration rates was closely tailed by a vast expansion of the penal system. Because prison capacity would lag always slightly behind the prison population, overcrowding became a perennial condition that demanded investment in ever larger and more numerous penitentiaries. The buildings were an expression of the structure of the system. Goldfarb despaired. He warned that bigger and better prisons would no more solve the incarceration problem than bigger and better roads would solve the traffic problem. Not only that, but market logic meant that if prisoner numbers continued to grow, penitentiaries could even become a lucrative business opportunity.

That warning was apparently read as a prescription: by the early 1980s, the penal system began outsourcing jail space to commercial providers. The Corrections Corporation of America was founded in 1983, and the following year opened its first private prison in Shelby County, Tennessee. Since then, the corporation, now known as CoreCivic, has only continued to expand. In 1997, they purchased the Central Treatment Facility in Washington, DC—a site formerly known as the DC Jail.

WHILE CALHOUN'S crowded rodents seemed to be so easily transferrable to the crowded human pens in newspaper stories, courtroom debates, and expert deliberations, psychologists argued that more research was required. Expert reports and scientific papers on prison environments often began with the observation that while the effect of crowding in animals had been "extensively examined," this was not the case with human populations. Paul Paulus, a psychologist at the University of Arlington, Texas, felts the connection between caged rats and human prisoners needed more investigation:

> All too often, the dramatic behavioral pathology of crowded rats described by Calhoun is used as a basis for predicting the responses of humans to conditions of high population concentration. While there is little doubt that humans will display behavioral pathology, if sufficiently crowded, we have no evidence indicating the kind of pathological changes that would occur or at what degree of crowding they would emerge.[17]

With Calhoun as an advisor, Paulus and his team at Arlington worked with Dallas County Jail and the Texarkana Federal Correctional Institution. Not only were they able to find crowded conditions that mimicked Calhoun's pens in being "intense, prolonged, inescapable, and realistic," they had a diversity of housing units to study—ranging from an open dormitory to the single cell.[18] They found close correlations between density and stress responses—leading to "increased suicides, psychiatric commitments, disciplinary infractions, violent deaths, and deaths due to natural causes."[19]

Just as Calhoun had said the Towson enclosure might have held five thousand rats if they were housed in individual cages, so Paulus's team recommended that the prison system provide more single cells if this prevented moving individuals into dormitories to meet the simplistic space-per-person standards now being requested of the correctional services.[20] They recognized that Calhoun's research was not simply about density in a physical sense but the number of individuals per square unit area. It was about degrees of social interaction.

Paulus and his colleagues participated in a series of well-publicized legal cases that helped establish changes to public health standards in prisons.[21] They openly expressed the significant role played by animal ecologists and ethologists in the development of their approach to the study of prison crowding, describing themselves as having been "stimulated by Calhoun's widely publicized studies with rodents showing a variety of deleterious effects of crowding."[22] Paulus's work reciprocally benefitted Calhoun, who, when approached by lawyers interested in the relevance of his work for other court cases and other prisons, was able to refer them to the ongoing studies at Texas that were now focused on crowding in human subjects, and on which he now advised.[23]

THE ECOLOGICAL or environmental approach was growing ever-more popular in psychology and psychiatry. The Environmental Design Research Association was founded in 1968, and the graduate training program in the new field of Environmental Psychology began at the Graduate Center of the City University New York. That same year, Esser had founded the Association for the Study of Man-Environment Relations and was editor of its journal, *Man-Environment Systems*.[24] By the early 1970s, this was joined by similar publications such as *Environment & Behavior* and *Design and Environment*.

Esser, Sommer, Izumi, and Calhoun were soon delivering lectures to architects and urban planners, encouraging then to think of the behavioral consequences of density and design. Environmental psychologists were advising on the design of homes for the aged, military facilities, college dormitories, schools, and airports, as well as hospitals and prisons. They were even beginning to advise on the design of the zoo; their studies of territoriality, space, and crowding, were beginning to flow full circle. The director of Atlanta Zoo, Terry Maple, named Sommer as a considerable influence, noting how zoo architecture was reminiscent of the prison—"hard on animals and the people who visited them."[25] For Maple, environmental psychology had an important effect on the zoo:

> Zoo architecture is now soft and naturalistic, and both the quality and the quantity of space is regarded

> as germane to the well-being of zoo animals and zoo visitors . . . Scientific zoos are no longer the exception; they are becoming the norm. At all levels, and around the world, science is alive and well at the zoo.[26]

The field was also beginning to outgrow the institutions to which psychologists had so neatly transferred the methods and insights from experiments with caged animals. Sommer and other environmental psychologists now wanted to transfer the ideas, methods, and concepts that they had developed in their experimental studies of crowded institutional environments to the world outside. Psychologists and psychiatrists fixed their attention on a larger prize—the city itself.

CHAPTER 13

THE RAT BILL

DELIVERING his first State of the Union address in January 1964, Lyndon B. Johnson laid out a bold vision for "The Great Society." Having accepted the office of president only seven weeks earlier, this was Johnson's first chance to talk policy to the American people, and he wanted to capture that aura of progressive optimism that JFK had first inspired and that his assassination latterly rendered quasi-mythical. In the spirit of FDR's New Deal, the Great Society would continue a tradition of placing public works ahead of international engagements, of improving the nation through community programs and investment in infrastructure. And so, even with the prospect of another foreign war building in Vietnam, Johnson declared a domestic War on Poverty. Converting that metaphor into political action would involve a series of programs variously addressing education, health, and housing. By mid-1967, most of the components had already been voted through in an omnibus bill when, on July 20, a final section reached the chamber for approval. On the schedule, this was listed as House Resolution 749, but everyone called it "the rat bill."

The Rat Extermination and Control Act of 1967 was to be part of the experimental Model Cities program, a flagship project of the recently formed Department of Housing and Urban Development, or HUD. Treating health, welfare, employment, and housing in unison, the Model Cities scheme was being rolled out over a hundred and fifty sites with a mission to improve conditions in the ghetto. The rat bill was an attempt to secure federal funding specifically to assist with rodent eradication. In wartime Baltimore, Curt Richter had been given twenty-five thousand dollars to clear the city of rats. To do the job nationally, the Johnson administration was asking Congress for forty million.

Rats, for Johnson, carried symbolic freight. To talk about rats was to conjure all the associated imagery of disease, degradation, and squalor. Added to which, rats were newsworthy—headline writers loved them. So by drawing special attention to the Rat Extermination and Control Act, Johnson figured the bill—and the rat—could act as a sort of mascot consolidating Model Cities, HUD, and his wider War on Poverty. All would not go to plan. "I thought the logic of exterminating rats was self-evident," he later reflected. "But I was wrong."[1]

The House found it ridiculous. To Johnson's dismay, the bill was met with roars of laughter. Congressmen called it "the civil *rats* bill," enquired whether the government was planning to create "a high commissioner of rats." Jokes bounced back and forth across the chamber, with Republicans and Southern Democrats taking aim. A congressman from Virginia played up his accent to announce: "Mr. Speaker, I think the 'rat smart thing' for us to do is to vote down this rat bill 'rat now.'" Part of their objection was the cost. Florida representative James Haley suggested it would be cheaper to "buy a lot of cats and turn them loose." Del Latta of Ohio could not understand why the administration needed a dedicated rat control program, or why it should cost so much: "Putting out a little rat poison is not too much to ask."[2] Clearly, neither congressman was acquainted with the work of David E. Davis.

The bill did not pass. With almost four hundred thousand US troops now on the ground in Vietnam and widespread civil unrest in the cities, this was hardly the time to be worrying about something as trivial as rat control. Compared to war, *The New York Times* suggested, the rat bill was a modest expense, and "would not pay for one month of the random cannon fire the armed forces lob into the Vietnamese jungle every night."[3] When the actual cost of the Vietnam conflict was later calculated, the *Times* estimate proved low. In fact, the war was mopping up the equivalent of the entire rat-bill budget every two days.[4]

Earlier that year, Martin Luther King, Jr., had already noted the disparity between government's rhetoric and its spending: "I knew America would never invest the necessary funds or energies in rehabilitation of its poor so long as adventures like Vietnam continued to draw men and skills and money like some demonic destructive suction tube."[5] War abroad became a pretext to deny help to the poor at home, and when it came to releasing funds, punishment was favored over prevention. Only the day before the rat bill was rejected, Congress had voted in favor of the Antiriot Act, which granted the federal government new powers to arrest anyone crossing state lines to "incite disturbances"—effectively, preventing demonstrators from travelling to national protests. "Yesterday you voted to establish federal supremacy to suppress violence," Theodore Kupferman, representative for New York's 17th District, told the House. "Today you voted to incite violence."

During that long, hot summer of 1967, more violence was a safe bet. In November 1963, tenants in Harlem, led by activist Jesse Gray, had organized a rent strike in protest at the dolorous conditions of their housing. The following summer, after a fifteen-year-old African American was shot by police, six nights of rioting engulfed the borough, leaving one dead and over a hundred injured. Since the riots in Harlem in July 1964, civil unrest had been escalating year on year. August 1965 saw the Watts riots in LA, where a violent arrest triggered five days of looting and arson during which the

National Guard was called out and thirty-four people died. Damage to property cost forty million dollars. In 1966, there were twenty-one major riots across the nation, and in 1967, over a hundred. What became known as the "Ghetto Riots" occurred all across America, in Louisville, Boston, Tampa, Cincinnati, Atlanta, Buffalo, and Toledo. On July 12, rioting in Newark left a further twenty-six dead. The spasms of urban unrest peaked later that month in Detroit. Three days after the rat bill had been voted down, police raided an unlicensed bar in the early hours of July 23. A bottle was thrown at the officers, the ensuing fight spilled out to the streets, and so began five days of the most destructive rioting since the Civil War. Ten thousand protestors burned more than four hundred buildings. With police and the National Guard overwhelmed, Johnson authorized the deployment of troops, sending in the 82nd and the 101st Airborne. There were tanks on the West Side streets, snipers mounted on rooftops. By its end, the Detroit uprising would see forty-three deaths, over a thousand injuries, and seven thousand arrests. That summer, *The New York Times* asked, "What is the Federal Government doing to prevent outbreaks of urban violence?"[6] and promptly answered that it had "not done enough."

In a second push to get his rat bill passed, Johnson now sought to draw a line directly from the control of rats to the control of riots: the waves of civil unrest sweeping America were a product of and reaction to ghetto life, the violence an inchoate protest against slum conditions. Casting opponents of his bill as opponents of good housing, Johnson blasted the vote as a "cruel blow to the poor children of America."[7] Rather than forget about the rats, he doubled down. "I spoke about rats in every public forum I could find," he recalled, framing the case to fit his audience:

> I argued economics with the conservatives: "if rats cost us about nine hundred million dollars a year, does it make economic sense to argue against a forty million dollar program of control?" I stressed morality with

> the moderates: "Have you ever lived in a broken-down tenement where you could hear rats scurrying inside the walls at night?"[8]

As with the Baltimore Plan, rat control was a synecdoche for urban decay, and support for the rat bill became a referendum on caring. In congressional debates, Democrats said their cities were reporting thousands of rat bites every year, citing the story an eight-month-old boy bitten to death in the nation's capital that very summer, of how the rat problem in the slums "haunts day-to-day existence." Describing it as "one of the most humane and compassionate bills ever to be considered by this body," William Barrett, a representative from Pennsylvania, scolded those who scoffed at the rat bill: "Let me assure my colleagues, Mr. Speaker, that in many of the areas of our cities this is no laughing or joking matter. Believe me, there is nothing funny about rats and rat bites." [9]

Johnson's ploy was working. Two weeks later, *The Evening Star* described the vote against the rat bill as "a Democratic public relations man's dream" and suggested that 1968 could be the "year of the rat." "Hardly a day goes by," wrote the *Star*, "without some Democrat denouncing the Republicans as being the party of the rats instead of the people."[10] The people took note. On August 7, seventy-five demonstrators, led by the Harlem rent-strike organizer Jesse Gray, forced their way into the House Galleries, chanting: "RATS CAUSE RIOTS!"[11]

FROM HIS lab at NIMH, Jack Calhoun was watching events in Washington closely. If the Johnson administration was going to do rat control, they had better do it right. A week before the rat bill went to the floor of the House, Calhoun had written to Johnson's Housing Secretary, Warren C. Weaver, to suggest the appointment of an advisory panel of scientists knowledgeable in "the ecology of rats and the social setting in which the activities of rats impinge upon human welfare." Calhoun volunteered his own services and recommended a list of rodent experts, including his former Baltimore colleagues John Christian (now at the Albert Einstein College of Medicine in New York), David Davis (recently appointed Chairman of the Department of Zoology at North Carolina State University), and John Emlen (now at Wisconsin-Madison):

> Any program of rat control which fails to tap into this reservoir of knowledge and experience stands a much greater chance of failure or misdirection of effort. Their addresses mark them as intellectuals, but all have intimate association with the real world of interface between rodent activity and human welfare. Their opinion will also not be marred with conflict of interest since they lack any direct link with commercial interests relating to rodent control.[12]

By "misdirection of effort" Calhoun chiefly meant poison. As the Rodent Ecology team had learned from Richter's efforts in Baltimore, while poisoning produced dramatic short-term results, the "only continuing effective control" was sufficient environmental modification as to make food and harborage unavailable. Fix that, Calhoun explained to Weaver, and all other means of rodent control became superfluous: "Any continued use of poisons and trapping in a particular locality merely indicates that ineffective environmental control has been conducted. Where environmental control has been effective there will remain so few rats as to preclude the necessity of poisoning or trapping."

When, at Calhoun's urging, Davis also wrote to Weaver the following week, it was to concur that poisons wouldn't work. By way of evidence, he referred Weaver to the impact of the Baltimore Sanitation Squads, which, by cleaning up alleys and repairing property, had brought the rat population of Baltimore from an estimated four hundred thousand in 1945 to a mere forty thousand in 1948. Davis bemoaned how, since then, the unstoppable popularity of warfarin—the miracle anticoagulant that replaced ANTU—had been worse for residents than it had for rats, simultaneously creating a "distraction from sound methods of housing improvement"[13] and allowing rat numbers to begin a steady return to their prewar figures. By 1952, amid widespread use of warfarin, Baltimore's rat population was over a hundred and fifty thousand, and its housing stock was falling back into disrepair. Consequently, both Calhoun and Davis stressed that HUD was precisely the place to base the rat program; Davis added that the rat bill transitively presented an "opportunity for permanent improvement of housing" because any commitment to long-term rat eradication also entailed long-term property maintenance. In terms of the material changes required, rat control and housing improvement involved identical steps—they were different ways to describe the same procedure.

Although both Calhoun and Davis endorsed the bill's potential to improve housing, they objected to Johnson's vilification of the rat. There was, the ecologists felt, really no need to stoke up negative feelings toward an already unpopular creature. Davis pointed out that rats were not the direct threat to public health they were once assumed to be and took pains to emphasize that "a rat is a symptom of bad housing, not a cause of bad housing." With rats high on the political agenda, Calhoun saw an opportunity to remind the public of their value and drafted a document entitled: "A Manifesto on Rats, Rodents and Human Welfare." Just as Davis had said the rat was symptom, not cause, so Calhoun warned that killing rats wouldn't automatically improve living conditions, and the monomaniacal focus on eradication was a misdirection:

> We might, with persisting intensive effort, remove the symbol. We might encourage several hundred thousand persons to set traps, we might supervise the distribution of

> tons of rat poison, we might organize watch-dog posses to ferret out each rat as some sign of it were noticed. All of this could be done—

—but we would never succeed in ridding the city of the rat, he said. Instead of trying to kill them, his manifesto urged, learning about how rats lived offered a way to better understand the effects of the urban environment on human residents. Calhoun emphasized the value of rodent studies, his own and others, for modelling the spectrum of psychopathologies generated by crowding and stress, and the strategies that might avert them. For Calhoun, it had been thinking about rodent control back in Baltimore that first steered him toward thinking of the implications for human societies of urbanization and high-density living. He now suggested that "we just might better quit looking directly at ourselves for a while, and instead look at rats to see what that avenue of searching tells us about ourselves."[14]

When Weaver replied, it was to thank them for the offer, but explain that the formation of an expert panel would be premature at this stage. For Calhoun, this was a disappointment. He had written to Weaver principally because he wanted to affirm that the expert view came down on the same side as the Johnson administration: bad housing led to bad behavior, and the link between rats and rioting really wasn't at all far-fetched. Rats were a sentinel species, their presence in any neighborhood "an indication of poor sanitation and housing conditions,"[15] as Davis put it. Calhoun and his Space Cadets had been building the case for over a decade: for incipient social unrest, the rat in the alley was the canary in the coal mine.

CHAPTER 14

SPACE CADETS

SINCE their formation in 1956, Calhoun and Duhl's Space Cadets had been discussing the link between housing and behavior. From his rat studies, Calhoun knew that manipulating the environment affected social behavior in rodents. He now wanted to canvass the experts he and Duhl had assembled to see how those effects might map onto human communities and what impact they might have on mental health and social order. As Calhoun explained to the group, the issue wasn't as simple as ridding the city of slums. They needed information to help them design and plan for the future, so as to build better homes for happy and healthy families.

The difficulty at this stage was the paucity of data about exactly what design changes would have which effects. During one of their first meetings, urban planner Melvin Webber had confessed that when making decisions about zoning or where to put a road, they were largely flying blind: "We don't know what the effect of the physical environment is on community values, certainly not on mental health. We are mostly riding on hunches, most of which I suspect are wrong."[1] Psychologist Daniel Wilner, who was engaged on a study of the psychological effects of overcrowding in downtown Baltimore, summed up Webber's problem neatly: "Planning is not well planned."[2]

Urbanist and systems theorist Richard Meier, who had been among the first to champion sustainability as a crucial part of city design, suggested it ought to be possible to find real-world cases where crowding in humans could be studied and the results compared with the extreme derangements seen in Calhoun's rats. "It's fairly obvious we wouldn't do Jack's experiment perhaps with humans," Meier said, "but we find society is doing them in many instances, and in a few instances the studies are being made on humans beings crowded together or being harassed for one reason or another, and we'd like to see what are the consequences."[3]

This was what had drawn the Space Cadets together: a desire to link up expertise from fields that were academically separated but addressing the same central problems. Calhoun asked the group to think about their own experiences of altering the urban landscape and learning from its effects on the population, to build a pool of case studies: "Each will serve as a sort of laboratory which can focus attention and focus our theoretical and practical questions on."[4]

WITH RAPID urban redevelopment underway across midcentury America, there was no shortage of natural experiments to complement Wilner's Baltimore study. Space Cadet regular John Seeley was director of the Alcohol Research Foundation, and over several years of discussion and argument with Calhoun and the others, he had become convinced of the value of developing an understanding of the environmental conditions that precipitated addiction.

The early 1950s had seen a marked change in the approach to addiction. In 1951, the recently formed World Health Organization convened an Expert Committee to investigate whether alcoholism should be classified as a disease. The WHO's motives were as much pecuniary as nosological: if alcoholism was a disease, it could be subsumed under the purview of a public health agenda and become eligible for funding accordingly. The WHO's mission was to persuade public health bodies that alcoholism fell in their court:

> If it can be shown that alcoholism as a behavior is *per se* a medical disorder and that socio-economic factors are contributing elements to its etiology, public-health workers will not shy away from it, as social and economic elements are involved in most or all health problems with which they have to cope.[5]

On the basis of the committee's report, the WHO declared alcoholism a disease in 1954. Two years later, the American Medical Association also agreed to classify alcoholism first as an "illness," later upgrading this to a "disease" in 1966.

Seeley wanted to advance the scientific understanding of alcoholism but felt the rush to categorise alcoholism as a disease had been premature: "It would seem to me infinitely preferable to say, 'It is best to look upon alcoholism as a disease *because* . . . ,' and to enumerate reasons."[6] Part of his reticence stemmed from a sense that the medicalization of alcoholism, while a useful means of removing stigma, also acted to minimize and conceal the social and economic factors that corresponded with addiction.

If the addictive properties of opioids were such that practically anyone could become addicted, alcohol was only addictive to certain people and under certain conditions. Few people tried heroin, but a significant proportion of those who did became addicted. A far larger number of people tried alcohol, but only a small proportion became addicted. Seeley suspected that alcohol was a drug whose misuse was likely to be an indicator of wider social problems. He set out to demonstrate that alcoholism could be mapped over other social and environmental variables, starting with population density. Taking his cue from the Space Cadet discussions of Calhoun's behavioral sinks and John Stewart's demographic gravitation, he sought to apply the principles of ecology and social physics to study the distribution of alcohol use and abuse across the United States. He called his project "The Ecology of Alcoholism."

Seeley first compared attitudes to alcohol—approval or disapproval—between states, and then between cities and rural areas within states. He discovered that people in cities generally had a more positive attitude to alcohol, whereas rural communities often favored a return to prohibition. Living in a city seemed to predict a more permissive attitude to drinking:

> We shall almost be driven to conclude that the population facts influence or determine the alcohol attitudes, since it is not reasonable to assume that alcohol attitudes determine the distribution of population over the United States. I say "almost" because we may also choose to assume that some common factor—such as a preference for city life or freedom from traditional restraints—underlies or influences both.

He next compared alcoholism rates and associated issues such as deaths from liver cirrhosis and found, again, that they varied according to population density—the denser an area, the higher the rates of alcoholism and its downstream pathologies.

To account for this, Seeley reached for the growing bestiary of animal studies—on deer, lemmings, and now also snowshoe hares—that had observed how elevated population density preceded sudden and massive die-offs. He didn't think human drunkenness would lead to human population collapse, but of particular interest was the condition of such animals prior to their deaths: enlarged adrenal glands, low blood-sugar levels, and liver cirrhosis; many of the physical symptoms of what Seeley still called "Selye's 'shock disease'" overlapped with those typically found in chronic alcoholics. But in addition, he noted that the crowded animals often behaved manically prior to death, swarming and panicking, displaying "excitement," "compulsive flight," and "blind rushing through what would normally be 'sources of satisfaction.'" Paired with the liver disease, this sort of reckless disregard for personal well-being exhibited by lemmings, snowshoe hares, and Calhoun's rats was, Seeley thought, "all too suggestive of behavior we encounter, suitably translated into human terms, in the alcoholic."[7]

Not only did drunkenness correlate with crowding, but crowded animals acted like they were drunk, abandoning the usual standards of restraint and engaging in higher levels of violence and sexual activity. Under conditions of inebriation or overcrowding, the response was the same: an inability to maintain the usual behavioral norms necessary for the successful functioning of social relations. If Seeley's intuition was correct, both crowding and drunkenness seemed to inhibit the performance of complex tasks, resulting in the expression of only the most basic behaviors.

SEELEY'S ECOLOGICAL mapping technique would be more thoroughly explored by another member of the Space Cadets, the Scottish landscape architect Ian McHarg. A former World War II paratrooper who had reached the rank of major and still wore a military moustache, McHarg was physically imposing, loquacious, charismatic. He was also a fervent advocate of Calhoun's work. During the early 1960s, McHarg had hosted his own television talk show series, *The House We Live In*, which pushed the ecological model of social analysis and whose guests included John Christian, Len Duhl, Lewis Mumford, Margaret Mead, and Hans Selye.[8] He had planned for Calhoun to appear, but their schedules never allowed it. When invited to a gathering of the Space Cadets in 1961, McHarg told the room:

> In all the years I have been studying city planning it never occurred to me the most important single statement about the physical behavior and about density was going to be said by Jack Christian and Dr. Calhoun. And they predicted the behavior of rats and density of social pressure and pathology, and profound analogies for city planning would come from this direction.[9]

A professor at the University of Pennsylvania's School of Architecture since 1954 and until his death in 2001, McHarg had developed a new form of "ecological planning" that sought to consider any new building project in concert with the existing natural functioning of the land. As laid out in a 1969 book, *Design With Nature*, his method was to mark geological and topographical data on a series of separate transparencies, then overlay these to plan a site for development. What he called his "layer-cake" technique was the precursor to what became geographical information systems, or GIS. The aim was to build sympathetically, to think about how humans could live within nature rather than on top of it.

When consulted on the proposed development of a rural valley just north of Baltimore in 1962, McHarg had persuaded the authorities to impose new segregated zoning ordinances. Where the segregation of 1911 had been to prevent racial mixing within the city, the new segregation was to protect the landscape from the spreading city itself. In Philadelphia, he had used his layer-cake method to chart the relationship between social pathology and the physical environment, overlaying on a map of the city the incidence of physical diseases, mental illness, violent crime, drug addiction, alcoholism, suicide, and pollution against factors such as employment, wealth, and population density. It was, in effect, Seeley's Ecology of Alcoholism but scaled up to include a much broader range of variables. Publishing the resulting heat maps side by side, McHarg explained the striking overlap of social density and social decay with reference to Calhoun's rat studies, concluding that—at least for humans in Philadelphia—"it seems clear that crowding, social pressure, and pathology do correlate."[10]

Both Seeley's "ecology of alcoholism" and McHarg's layer-cake maps provided a strong indication that social pathologies such as crime and addiction occurred coincident with population density. This was all good empirical data, but the studies—composed from statistical information—were short on detail and said little about how those factors played out on the ground. As it happened, Len Duhl had recently arranged for NIMH to fund a project based at Boston's Massachusetts General Hospital and overseen by two other Space Cadet members, psychiatrist Erich Lindemann and social psychologist Marc Fried.

IN BOSTON, construction had begun on a new highway in 1953. The six-lane Central Artery would displace more than twenty thousand residents, almost all working class, destroying much of the city's historic West End and Chinatown. Shortly before the highway was due to open in 1959, the Boston Redevelopment Authority decided to raze an additional forty-eight-acre portion of the old West End alongside the Charles River. Demolition began in 1958, wrecking balls brought down nine hundred buildings, and bulldozers scraped the land flat. The Catholic Church, St. Joseph's, was spared. Almost three thousand families were displaced to make way for a suite of high-rise and high-rent luxury apartment complexes.

The levelling of the old West End was justified by the assessment metrics used by the Boston Housing Authority, according to which the district ticked all the boxes: it was "over-populated," "densely covered," "overcrowded," and suffering from a "severe lack of any open space." It was, officially, a slum.[11] By contrast, the new Charles River Park apartments advertised themselves in terms of their spaciousness and privacy.[12] Although former residents were invited to move in to the new buildings, the redevelopment was a de facto evacuation. With rent prices for the new apartments set at ten times what they had been paying, the citizens of the old West End were scattered. "From a social aspect, it was a tragedy," recalled actor Leonard Nimoy, who had grown up on Chambers Street, "because a wonderful, tight community was destroyed."[13]

With the mass evictions planned right next door to their offices at Massachusetts General Hospital, Lindemann and Fried saw a chance to investigate the effect of the disruption to social life and mental health in the West End communities, an opportunity "to question the extent to which, through urban renewal, we relieve a situation of stress or create further damage."[14] Fried had come to believe that the working classes experienced the urban environment very differently from the middle classes, with a more permeable boundary between home and street. Where the middle classes might draw a strict perimeter around the property they owned—the white picket fence, the concierge-accessed lobby—the home of the West End tenant frequently bulged out into street, running over the pavement or porch steps or shop front. Fried had found that the work of ecologists and ethologists had "considerable bearing" on his studies of how the working class shared common spaces and spoke of these in terms of territoriality and home range. With so little truly private space, the social relations between residents remained harmonious through a foam of interlocking territorial domains, intimately mapped over and embedded within the physical environment. Consequently, precisely because the population density was so high, the social relations between residents and their senses of self-identity were more tightly bound up with the specific structure of a particular neighborhood, making relocation especially injurious.

Unravelling the dense tangle of social and spatial relations was going to be difficult, but help came in the form of a thirty-year-old sociologist, Herbert Gans, whom Duhl had introduced to Fried. Gans's approach would be immersive: "I wanted to know what a slum was like," he said, "and how it felt to live in one."[15] So, in October 1957, Gans and his wife moved in to a small apartment in Boston's West End.

They spent almost a year living in a largely Italian American enclave, Gans keeping a field diary monitoring the movements and interactions between his neighbors. During the study, Duhl invited Fried and Gans to speak to the Space Cadets, where Gans explained his methodology:

> I am trying hard to observe deliberately all those things one takes for granted in city living. For example, it is very interesting to see how people talk to each other on the street, how they use the street, when they gossip in front of the house, and when they talk on street corners. Then there are a series of events that one observes just by living in the area and being around at all times.[16]

The research wasn't all so furtive. The Ganses tried to get to know as many families as they could and regularly attended local dances and events. "Herb's expense account included beer at the local pubs," Fried reported, "where he spent a great deal of time talking with the late-adolescents."[17] "I probably look several years younger than I actually am, so I can pass as a student and thereby place myself in a better status of position to obtain information," Gans explained, "and if they discover me to be a scientist, it's not as if I were posing to be someone entirely different."[18]

Where the Boston Housing Authority had applied its rigid metrics to designate the whole of the West End as a slum, Gans found instead a network of separate subcommunities, each with its own social system and sense of identity. As he told the Space Cadets in 1957: "The buildings may look poor on the outside, but inside there are perfectly respectable apartments that are kept up in ways not too different from middle class standards." While there were certainly parts of the West End that were filthy and run-down, the blanket categorization missed the differences between and even within blocks. "When people in the project area were told they were located in a slum area," Gans noted, "many of them wouldn't believe it."

Herb Gans's longstanding concern was that the people making decisions about redevelopment were really making lifestyle judgments. City planners and civic authorities looked at the low-income communities "act on the assumption that this way of life is simply a deviant form of the dominant American middle-class one."[19] Fried argued that in light of the evidence from the West End, they ought to think carefully about how crowding and density was defined and understood. Rather than helping poor communities, the equation of crowding and behavioral decay was being used by to justify slum clearances. "Until we can prove these patterns are *pathological* rather than different," Gans argued, "I don't think we have any right in our planning to impose anything."[20]

IN 1962, Gans collected his work from Boston into a book. By the time *The Urban Villagers* was published, it was too late to save the West End: the communities had long since been moved out, and the luxury new high-rises were complete. But for architectural journalist Jane Jacobs—who was at that time battling to save her own urban village from a Robert Moses–sponsored redevelopment plan—Gans and Fried's account of rootedness provided vital evidence substantiating her defense of old New York. As destructive clearance programs spread out across American cities, "Remember the West End!" would become a rallying cry for those seeking to defend their communities from the bulldozers.

Jacobs was an advocate of busy places. She felt the planners tasked with redesigning America's cities had misunderstood what she called "the remarkable intricacy and liveliness of downtown"[21] as mere chaos and disorder. Consequently, they sought to enforce orderliness, and in so doing sapped the vitality of the urban streets. "From city to city the architects' sketches conjure up the same dreary scene," Jacobs lamented. "These projects will not revitalize downtown; they will deaden it."[22]

Her longtime champion and collaborator William "Holly" Whyte was aligned. Whyte had made his name in the early 1950s with a series of influential books and articles that examined the peculiar dissonance between the American rhetoric around individualism and a culture which rewarded compliance and uniformity. People talked about their freedom but acted like drones. Following George Orwell's template for "newspeak" from the recently published *Nineteen Eighty-Four*, Whyte called that conformity *groupthink*, and later typified the compliant worker as "The Organization Man."[23] In the growing suburbs, Whyte saw the groupthink of his Organization Man manifested in row upon row of identical houses. As Whyte saw things, "the norm of American aspiration is now in suburbia. The happy family of TV commercials, of magazine covers and ads, lives in suburbia."[24]

By contrast, the city was a site of creativity and spontaneity. The bustle and press of the streets weren't stressful but invigorating. It was precisely the likelihood of encountering strangers—Whyte called it propinquity—that made the city valuable. Cities were messy, and that was a positive, something that needed to be preserved in their redevelopment. As Jacobs put it: "If this means leaving room for the incongruous, or the vulgar or the strange, that is part of the challenge, not the problem."[25] Like Jacobs, Whyte saw the new redevelopments as sterile and oppressive: "They are the concrete manifestation—and how literally—of a deep, and at times arrogant, misunderstanding of the function of the city."

Whyte came to believe that cities were being remade by people who simply didn't like cities. He later recalled the impact Calhoun's work had on the discussions around city living and that when he began studying urban behavior, the "spectre of overcrowding was a popular worry. High density was under attack as a major social ill and so was the city itself. 'Behavioral sink' was the new pejorative."[26] But in Whyte's view, there was a strong case to be made for the benefits of crowding:

> There is a rash of studies underway designed to uncover the bad consequences of overcrowding. This is all very well as far as it goes, but it only goes in one direction. What about undercrowding? The researchers would be a lot more objective if they paid as much attention to the possible effects on people of relative isolation and lack of propinquity. Maybe some of those rats they study get lonely too.[27]

Whyte was very taken with Gans's methodology for people-watching in Boston and would later embark on his own similar project in New York. He contrasted rat studies with his own people-entered approach, complaining that Calhoun's "research was vicarious; it was once or twice removed from the ultimate reality being studied. That reality was people in everyday situations. That is what we studied." Across New York City, on street corners and public squares, in plazas and doorways, his team set up cameras to observe how pedestrians behaved. Whyte's philosophy: "Schmoozers are instructive to watch." He compiled hundreds of hours of footage of the interactions between strangers or colleagues meeting for lunch; how people avoided bumping into one another on busy pavements, how they reacted when they did; how groups of men watched women walking by but never spoke to any; how the drunk or the homeless were given wide berth or small change.

The Street Life Project would eventually lead to a popular 1980 documentary film, *The Social Life of Small Spaces*. Much like Hall's style of anthropology, Whyte's applied wry observational wit to describe the spontaneous choreography of pedestrian motion. He noticed that rather than avoid contact, people routinely placed themselves in one another's way—sitting on steps, standing in doorways. He noticed that "people who stop to talk gravitate to the center of the pedestrian traffic stream." This was the urban dance, and density was not the problem, it was the solution. Whyte's prescription: in order to save the city, architects must avoid wide empty plazas, which remained desolate and unused, and instead design spaces that increased human clutter:

> The approaches that work the best are those which meet the city on its own gritty terms; which raise the density, rather than lower it; which concentrate, tightening up the fabric, and get the pedestrian back on the street.

Unlike Calhoun, who had pathologized the draw of the crowd, Whyte celebrated the social vortex of urban spaces as "manifestations of one of the most powerful of impulses: the impulse to the center."[28]

AMONG CALHOUN'S circle of experts, a schism was developing. Population density seemed to be correlated with higher rates of alcoholism, crime, violence, and sexual deviancy—all of which were in line with the rats of Casey's barn, and all of which seemed to condemn the city. Wilner's work in Baltimore had provided persuasive empirical evidence that more spacious accommodation seemingly led to better life outcomes across a broad range of indicators, while Seeley and McHarg both found persistent overlaps between population density and social pathology. And yet, cheering on the cities were the likes of Herb Gans, Jane Jacobs, and Holly Whyte—all of whom saw in the high-density urban environment a socially complex and heterogenous network of communities. In addition, Gans and Fried had made the case that working-class areas were being denigrated as slums and slated for demolition simply because they didn't look or function like middle-class areas—that what was really being enacted was the removal of a way of life that wealthy elites found vulgar and unseemly.

McHarg, who had grown up near the slums of prewar Glasgow, flipped that class-based argument on its head. He had little sympathy for the idea that there was anything picturesque about poverty. He recalled his 1930s childhood in characteristically trenchant terms:

> Almost ten miles from my home lay the city of Glasgow, one of the most implacable testaments to the city of toil in all of christendom, a memorial to an inordinate capacity to create ugliness, a sandstone excretion cemented with smoke and grime. . . . much of the city was a no-place, despondent, dreary beyond description, grimy, gritty, squalid, enduringly ugly and dispiriting.[29]

Like Holly Whyte, McHarg had also made a film about cities. But where Whyte's avuncular narration had cheerfully described the foibles and mores of New York's street life in *The Social Life of Small Spaces*, McHarg's film took an entirely more sinister view of urban life. Released in 1969, *Multiply and Subdue the Earth* was apocalyptic. A cold open showed footage of the rats from Casey's barn; the voice-over is Jack Calhoun's, his measured Southern delivery carefully enunciating the findings of his studies. As Calhoun describes how "a larger and larger proportion of the population displayed deviant behavior that was not of survival value to the group"[30] and goes on to describe the maternal neglect and homosexuality witnessed during the behavioral sink, the film cuts back and forth between the rats and scenes from the seedier side of 1960s New York: drunks in bars, porno theatres on 42nd Street. A swarm of fighting rats cuts to the tide of commuters emerging from a subway station. McHarg was not a cheerleader for the modern city.

Now confronted with the idea that working-class communities had developed a convenient fondness for the conditions of high social density in which they were housed, McHarg balked. There was, he told a conference in Virginia, a biological limit to adaptation. Crowding increased pollution, noise, and overstimulation—all of which were intolerable. To the idea that the poor enjoyed their lot, McHarg, in his rich brogue, thundered: "No, we say overcrowding is fine; the working class like it. I'll be damned if the working class like it."[31]

HOW, THEN, to reconcile these two accounts? The solution lay in a more precise account of crowding. When measuring density, planners were often using a crude measure of people per unit area. But, as Gans pointed out, "you can have a thousand people per acre in the city, but if they live three to a room, I think they will be a hundred times as much trouble as if they live one to a room." The population density on wealthy Fifth Avenue was very similar to that on the poorer Lower East Side, but the arrangement of the space, the distribution of the population within it, and the ability of the residents to travel elsewhere all affected their susceptibility to crowding stresses.

Affluent people experienced space differently because they had the means to escape and access to privacy. Free to travel more widely, they were less confined to local neighborhoods. Even within their neighborhoods, they could retreat to the seclusion of a spacious apartment. Crowding was to be distinguished from overcrowding, which was most certainly pathological and experienced more frequently by the poor and the marginalized. Ned Hall put it sharply: "Education and position on the socio-economic scale seem to effect vulnerability to crowding. Those that are at the lower limits are apparently more vulnerable."[32] Jane Jacobs made much the same point when she argued that measures of density captured only the numbers of dwellings per acre of land, not their arrangement. Density could be positive; *over*crowding—which meant "too many people in a dwelling for the number of rooms it contains"—never was. Too often, high density and overcrowding were seen a synonymous, coupled together, she complained, "like ham and eggs, so that to this day housers and planners pop out the phrase as if it were one word, 'highdensityandovercrowding.'"[33]

Critical was the frequency of social contact, something that Calhoun had considered very carefully. Measuring the frequency of social contact in relation to social status among his rats, he had devised a useful concept—"social velocity." Social velocity, which he also described as social temperature, was a measure of how frequent and how involved each individual's social interactions were. The concept drew upon his understanding of an ideal group size among both rats and humans—between eight and sixteen adults, with an optimal or basic group size of twelve adults. This was an evolutionary template, he argued, as our primate ancestors struggled to survive as semi-isolated groups: "More recent cultural evolutionary merely overlies this primitive genetic base."[34]

Among both rats and humans, the optimum group size facilitated an individual's optimal social and psychological state: too small a group was under stimulating; too large, and the individual would become overwhelmed by unwanted interaction, leading to frustration and eventually to violence and withdrawal. In order to cope with excessive interactions, what Calhoun called the "intensity" of each interaction would be reduced—"up to the point where the intensity of interaction becomes so shallow as to convey no meaning."[35] As the rat populations grew in a pen, subclasses of individuals emerged that could be classified according to their velocity. The higher-status animals tended to be more active, able to move around the pen more freely and have more rewarding social contacts. They were easily distinguished from their low-velocity and lower-status equivalents, who were socially isolated and physically immobile.[36]

Discussing how his rodent universes were relevant for human environments, Calhoun noted that household overcrowding in humans and pen overcrowding in rats illustrated how physical and social variables were closely entwined. "I realize that the physical situation has no reality without considering the social organization," he conceded, "but neither can we just consider social organization because the social organization has no reality in many instances without physical situations."[37] Much as Osmond and Sommer had found at Weyburn, or Paulus in the Texas prisons, it wasn't simply an issue of density per unit area, but the design and use of adequate social spaces within that area.

The key factor seemed to be some form of personal living space, as Richard Meier explained: "The experiments in the laboratory suggest that *privacy*, as much as it can exist for animals, seems to be essential for community peace."[38] However, as the new housing developments began to sprout over the cleared slums, privacy was a commodity in desperately short supply.

IN 1963, Duhl invited thirty-one of the experts who had presented at Space Cadet meetings over the years to contribute to an edited collection, titled *The Urban Condition*. Marc Fried, Herb Gans, Ian McHarg, Richard Meier, John Seeley, and Daniel Wilner are all there. Calhoun's entry is a reprint of his *Scientific American* article, "Population Density and Social Pathology." Robert C. Weaver, later the Housing Secretary under President Johnson, is also included. That there were almost as many disciplinary backgrounds as there were contributors was crucial. "Eight years ago," Duhl explains in his introduction, "my own interest in the relationship of these many disciplines to mental health, along with John Calhoun's concern with the impact of the physical environment upon behavior, led to the creation of this group."[39] Duhl notes that the broad array of approaches are held together by a common concern with processes and systems: "The 'models' employed in this book are several. The one that seems to predominate is that which the biologists call 'ecology.'"

The following year, as part of his War on Poverty, President Johnson established a Task Force on Urban Problems, chaired by Robert C. Wood, that would eventually lead to the founding of HUD. Upon hearing of this, Duhl submitted a two-page memorandum to the committee, pleading that when considering those urban problems they privilege the "social and psychological" over "bricks and mortar."[40] Duhl had a practical suggestion, too: they should select a limited number of urban sites in which the sort of integrated approach taken by the Space Cadets could be demonstrated in the field, and "in which we could show that by doing planning and policy holistically, we would do better."[41] Duhl suggests they call these "Demonstration Cities." Wood liked the idea and credited it to "Duhl's own advisory group (dubbed 'space cadets')."[42] When the proposal for Demonstration Cities reached the president, Duhl was later told that Johnson had said: "I've had enough demonstrations. Let's call it Model Cities."[43]

Wood was subsequently promoted to HUD undersecretary in 1965 and, when the Model Cities program was announced, offered Duhl a role as a special consultant. In 1966, Duhl—whose early experience volunteering for the Public Health Service had first turned his attention to the importance of community structure to mental health—left NIMH to take a position in Washington. In the Model Cities program, he saw a chance to put into practice what he and Calhoun had been thinking about for the past twelve years. Duhl's departure would signal the end of the Space Cadets, and they held their final meeting that same year.

Announcing his new role at the Space Cadet meeting of March 1966, Duhl spoke enthusiastically of the possibilities at HUD for developing a "total program" that recognized the urban complex in its entirety, moving beyond the fragmented approach of multiple government agencies each focused on "tiny segments of the city." He described how he had become increasingly frustrated at NIMH, where the broad remit of the Institute meant his attention was spread thin—flitting from mental disability, to alcoholism, to Boston's West End, to the poverty program. The move to HUD meant more focus, and the chance to really make a difference. Although leaving NIMH meant the end for the Space Cadets, he told the group that he hoped they would be able to continue their "informal arrangements among ourselves" and that by "operating in many different separate parts, we can push in this same general direction."

Duhl had only been in his new job for three days, but was already able to report progress: his "Demonstration Cities" proposal—"an attempt to give HUD the possibility of coherently pulling together programs from almost all of the agencies in government"—was encouraging the first coherent plans integrating the design of physical space with health and social planning, and involving local residents in the process. Duhl felt that all they had been talking about over the years was finally happening, and he was clearly impressed by the changing ethos and style of federal government administration: "The whole place is being rocked, as you can imagine, by the changes." He concluded the meeting that day with a message of hope, declaring it the Space Cadet's "last meeting and the HUD-nuts first meeting."[44]

Everything seemed to be coming together. Even the provision for rat control was eventually passed the following year, on September 20, 1967, as part of the Partnership for Health Amendment Act. That December, Johnson proudly declared, "We passed that rat bill because a Nation's conscience cried out louder than Republican laughter."[45]

CHAPTER 15

VERTICAL SLUMS

AN article in The Architectural Forum in April 1951 had nothing but praise for a new development just breaking ground in Missouri. It was to cost fifty-eight million dollars, housing fifteen thousand residents. Orderly and sanitary modern structures would replace "ramshackle old houses jammed with people—and rats."[1] The report made the cover story, titled "Slum Surgery in St Louis." Like most housing projects of the era, the blocks were to be racially segregated, with black residents in a section named for African American war hero and Tuskegee Airman Wendell O. Pruitt, and white residents in buildings named for progressive Missourian politician William L. Igoe. Hence, even the name was segregated: Pruitt-Igoe.

The lead architect was Seattle-born Minoru Yamasaki, who would later design the World Trade Center in New York. Rather than spread buildings low over the site, he concentrated the required quota of almost three thousand units into thirty-three eleven-story buildings, leaving plenty of open space at ground level for playgrounds and parks. The plans envisaged rivers of trees between the rows of accommodation. Budgetary constraints meant the rooms within the apartments were quite compact, so Yamasaki compensated with wider corridors and balconies. *The Architectural Record* put a positive spin on this, claiming that the design achieved "that essential smallness of scale within the huge context of the project which alone will preserve conditions in which human beings can live comfortably and retain all that is possible of the small neighborhood."[2]

The result looked more like a series of warehouse racks on a concrete floor. The rivers of trees were never planted, and even the playgrounds were only added after residents complained. Insufficient funds had been set aside for maintenance, and disrepair accumulated. Although the site was large, the buildings themselves felt crowded and claustrophobic. The elevators—which by design only stopped at every third floor—were often faulty, forcing residents to cluster in the narrow stairwells. Crime began to rise.

As architect and city planner Oscar Newman explained when interviewed for a BBC documentary at the site in 1974, "Gangs began to move in and occupy the apartments, and use the apartments as a base of operations to victimize the rest of the population." Meanwhile, the playgrounds and open spaces between the buildings, too far below to be available for unsupervised play, "became sewers of glass and garbage, rather than rivers of trees."[3] The ground floor became a wilderness, roamed by gangs. Windows were broken so frequently repair became impossible. By 1968, with over ten thousand missing panes of glass, vacant apartments and corridors were unprotected from the elements. Rain blew in. When the temperature dropped below freezing that November, water pipes throughout the blocks burst, flooding apartments as the thaw came. People began moving out. The buildings had never achieved full occupancy, but vacancy rates now fell to fifty percent. Then eighty-five percent.

By the time Newman was interviewed, Pruitt-Igoe was a wasteland of rubble, the first towers spectacularly brought down with controlled demolitions broadcast live on television in March of 1972. Unintentionally echoing the language once used by *The Architectural Forum*, the St. Louis

Housing Authority called the operation a "surgical demolition."[4] Filmmaker Ron Fricke collected helicopter footage to use for the climactic scenes of Godrey Reggio's *Koyaanisqatsi*, over which composer Philip Glass, on his first film commission, scored pulsing arpeggios to accompany Pruitt-Igoe's dramatic collapse.

Retrospectively calling Pruitt-Igoe a "prescription for disaster," *The New York Times* commended the decision to demolish the site:

> When officials dynamited Pruitt-Igoe Houses in St. Louis this year, they finally blasted the subject of housing design into the public consciousness. It took the violent and necessary act of destruction of part of a public housing project that had become an obscenity of American life to make it clear that we have been doing something awfully wrong.[5]

This sense of failure was focused again on the problem of design, rather than the continued problems of poverty and inequality. The architectural historian Charles Jencks later declared, "Modern Architecture died in St. Louis, Missouri on July 15, 1972, at 3.32pm (or thereabouts) when the infamous Pruitt Igoe scheme, or rather several of its slab blocks, were given the final coup de grâce by dynamite."[6] Similarly, *The New York Times*: "Le Corbusier's *Ville Radieuse* and dreams of the modern movement, R.I.P."[7]

Along with sounding the death knell for Modernism, Pruitt-Igoe also became an emblem of all that was wrong with public housing. Notable was the extent to which the physical environment was blamed for incubating the crime and disorder that ensued. Predictably, the rat-crowding studies were invoked to explain this, with sociologist A. R. Gillis describing how "social conditions in the project deteriorated to a level close to Calhoun's description of a 'behavioral sink.'" Because boxers Michael and Leon Spinks, the first brothers to both win gold medals at the same Olympics and in the same sport, had spent their early childhood in Pruitt-Igoe, Gillis derisively added: "The level of disorganization and conflict was sufficient to produce the Spinks brothers, outstanding in their boxing skills and motor offences."[8]

WHEN CALHOUN himself looked at the high-rise new projects, they reminded him of the highly organized and industrialized laboratories he had seen at Johns Hopkins and Walter Reed—geometrical arrays of identical units such as Richter's regimented rat rooms or Joe Brady's metal crates—in which a single individual was kept isolated from the stress of social contact. With one significant departure: in the apartment building, the inhabitants had to leave their cages to source food and other essentials. The stairwells of housing projects such as Pruitt-Igoe had influenced his design of the elevated burrows in Casey's barn. When Calhoun had explained to the Space Cadets how he had placed the burrows "several feet up and the rats have to run up a spiral staircase,"[9] Daniel Wilner had picked up on the comparison:

> The Baltimore slum is perhaps—at most—three story, half Civil War buildings, which are very old indeed. But how many children are you likely to encounter in the hall in any event? Pretty small. When you crowd 20 children into a single elevator it is a different story.[10]

Those allies of Calhoun who had studied the relationship between space and behavior in institutions such as the mental hospital and the prison found it disturbingly easy to transfer their findings to the semi-institutional environment of the housing project, with its similarly contained populations and predictable patterns of activity. Architect Kiyoshi Izumi, who had once tripped down the corridors of Weyburn with Humphry Osmond, was now working as an advisor to the Canadian Ministry of State for Urban Affairs. He predicted dire consequences should city planners neglect the physical design of spaces while imposing rigid codes and standards regarding density—be they "numbers of people or housing units per acre, numbers of patients on a ward, numbers of pupils in a classroom and so on."[11] Drawing from both Calhoun and Hall, he warned that the increased crowding and overconcentration of people in buildings that did not provide for privacy and community would result in residents becoming isolated, alienated, and institutionalized.

When, in 1967, Ned Hall had taken a journalist to Chicago's Cabrini-Green estate—a housing development that vied with Pruitt-Igoe for notoriety—he explained to his interviewer how the high-rise "vertical ghetto" was designed for everyone except its residents:

> The thing about a slum is that you can see it. The reason for this particular solution is that it walls off what's behind it. Slums, poor people, dirt and filth—they're not attractive. People don't like to look at them. So you build one of these things and the problem becomes invisible.[12]

High-rise tower blocks provided an only cosmetic fix; as Hall put it, "less distressing to look at than slums but more disturbing to live in."[13] Although the buildings seemed orderly on the architectural renderings, the result was a spatial environment that was disruptive and bewildering for its inhabitants.

Now widely recognized as an expert on space and design, Hall, along with his wife Mildred, was employed by the City of St. Louis on a Pruitt-Igoe Action Team in 1971. It was a last-ditch attempt to try to save a project now "known all over the world," as they described it, "as a disaster."[14] Following a visit to Pruitt-Igoe in the summer, Hall was sufficiently disturbed to write to a friend in Pittsburgh that the place was "beyond belief man. I stayed awake all night thinking about it. . . . The things they told us would blow your mind."[15]

When sociologist Lee Rainwater visited Pruitt-Igoe in 1966, it had struck him that the buildings failed to provide residents any sense of security, what he termed "the house-as-haven." Their surroundings offered nothing to be proud of, nor any sense of refuge. The already diminished self-esteem of the residents was reflected back at them by the disrepair of their housing. A destructive feedback loop ensued: "The physical evidence of trash, poor plumbing and the stink that goes with it, rats and other vermin, deepens their feeling of being moral outcasts. Their physical world is telling them that they are inferior and bad just as effectively perhaps as do their human interactions."[16]

Rainwater's observation that the buildings deprived their inhabitants of both privacy and a sense of ownership had resonated with Oscar Newman, the architect interviewed by the BBC shortly after the demolition of Pruitt-Igoe. Newman had first visited Pruitt-Igoe in 1964, part of a team of architects and sociologists employed to study the site. As he later explained to *Time* magazine:

> Every public area—the lobbies, the laundries and the mail rooms—was a mess, literally. There was human excrement in the halls. Except on one small area on each floor of each building . . . a little hallway separating two apartments. This little hall was spotless—you could eat off the floor. When we called out to each other in the other hallways, we could hear people bolting and chaining their doors, but in this area we heard peepholes click open. Sometimes people even opened their doors. The reason was that they felt this little hallway was an extension of their own apartments. We knew we were onto something.[17]

Now with Rainwater's account of the house-as-haven in mind, he realized what that something was. What people sought from any shelter was protection; a house was a territory, and a territory must be defended. The inhabitants of Pruitt-Igoe felt neither the desire nor the ability to defend where they lived: they lacked a territory. Drawing on the ecological concept of territoriality, Newman called his new theory "defensible space."

In Boston's West End, Gans had noticed how the domain of the family home spilled out to include a portion of sidewalk, or a stoop. A tall building made this sort of spillover impossible: "The only 'defensible' space becomes the apartment itself. The blind elevators, the long, anonymous double-loaded corridors, and the enclosed fire stairs are a no-man's land made to order for anti-social activity."[18] Deprived of any sense of territorial possession, the residents grew indifferent and eventually hostile toward their surroundings. It should come as no surprise, Newman argued, that they subsequently "treat their dwellings as prisoners treat the institutions in which they are housed."[19] The self-inflicted vandalism of public housing was a prison riot in slow motion.

For Newman, the timing of Pruitt-Igoe's demolition was perfect: in 1972, he published *Defensible Space: Crime Prevention through Urban Design*, which became an influential text among both designers and policymakers. In it, he promoted low-rise, low-density developments, particularly for multi-family dwellings, while damning the "larger-and-larger cookie cutter formula projects of clustered high-rise buildings on superblocks of open space . . . the guaranteed prescription for disaster."

For all that it seemed inspired by the same ecological turn that had animated the Space Cadets, Newman's account of territoriality was rather hazy. He never explicitly mentions any animal models of territoriality, but instead refers his readers to Robert Ardrey—whose books had by now made the concept central and familiar. Like Ardrey, Newman held that territoriality and violence were innate and universal. Crucially, his book suggested the Modernists had been wrong: people couldn't live within machines, and a new architectural style could not supplant natural patterns of behavior. Newman claimed nature and history were on his side: "The evolution of human habitat over the past thousands of years" had always involved the building of structures that defined "the territorial realm of their dwellings." And he lamented how, during a time of unprecedentedly rapid growth of American cities, that "tradition, grown over thousands of years in man's piecemeal search for a form of residence in an urban setting, has been lost."

THE CONTINUITIES between Hall's and Newman's studies of public housing and Calhoun's rodent pens were striking. In the enclosures at Casey's barn, Calhoun noted how the ramps allowed for the successful defense of the end cells, where mothers successfully raised their young. In pens where there was no such protection or refuge, crowding increased, and social disorder and violence ensued. The partitions, elevated burrows, and dividing ramps had all been means of tampering with the territorial impulses of his rats, assuaging or aggravating social strife between them.

Calhoun's rat pens remained a useful way of explaining the problems of planning the urban environment, and what he and his fellow Space Cadets had achieved was very significant. While they had provided a powerful critique of existing programs of slum clearance and urban renewal, they also emphasized the importance of the physical environment for mental health through designing more effectively for both community and privacy. As Calhoun reflected in a report for his superiors:

> Although, it is difficult to assess any one individual's role in these conferences, it is undoubtedly true that the informal communication channels opened through the medium of these conferences has led to a wide dissemination of Dr. Calhoun's concepts and philosophy of viewing the origin of an individual's behavior as a consequence of the milieu in which it lives.[20]

His studies had also provided them with a warning of how good intentions in design could go badly wrong. The early optimism surrounding public housing was something that Calhoun recognized, and he described to the Space Cadets how he also had fallen into the same trap at the very start of his studies in Casey's barn:

> **CALHOUN:** I thought this was an ideal environment for rats.
> **DUHL:** There go your middle-class values again. (Laughter)
> **CALHOUN:** I lived in the rat slums where conditions were terrible with wild rats, and I felt if I built a situation and kept it at a certain density it would be ideal. I made all sorts of calculations of mortality and death rates so I could predict populations and remove animals at different times, and I expected an equal distribution of animals more or less.
> **DUHL:** You just can't play God, Jack. (Laughter)
> **CALHOUN:** Even with rats.[21]

He explained how he came to realize that it hadn't been the size so much as *distribution* of the population that was the problem; even when numbers were held stable at approximately eighty adults, their unbalanced organization in the pens had resulted in the behavioral sink.

However, while Newman stopped with spatial design and a focus on territorial defense, Hall and Calhoun and Duhl insisted on integrating design modifications with associated community programs. Through the Space Cadets and its portfolio of urban laboratories, both human and animal, they had provided a more sophisticated account of how physical environment shaped behavior and how that was enmeshed with the frequency of interactions and social hierarchy. Consequently, when Ned Hall spoke of the problem of Pruitt-Igoe, he went beyond its physical design, arguing that public housing was a "social machine" as well as a technological innovation. To work effectively it needed to employ members of the community to work together in its management, protection, and maintenance, and to attend to issues of employment, transportation, stores, and social service programs.[22] As he explained to a friend: "The people to whom we are consulting are now convinced that unless you have a complete program (jobs, child care, schools, shops, transportation, and security) anything else in the way of housing is futile."[23]

As important as it was, Hall did not believe bad design alone was a sufficient explanation for urban decay, or, accordingly, that good design would be a sufficient solution. It was a view shared by the legal scholar Lawrence Friedman, who offered a qualified defense of the housing programs. Friedman saw public housing as having provided many concrete improvements to the lives of their residents: "In the public housing projects of most cities, mothers sleep without fear of fire, and rats do not bite sleeping babies." Public housing needed reform, not abandonment, and he decried the continued problems of segregation and economic inequality in American society: "Public housing has committed many crimes, no doubt; but ironically its conviction and sentence stem from crimes it did not commit and conditions over which it as little or no control."[24]

Hall's concern with the social as well as the physical environment allied him with Len Duhl, who had left NIMH for HUD's Model Cities program with high hopes. The Model Cities program was supposed to help resolve some of these wider social and economic issues that plagued programs such as public housing. But Duhl's optimism had been short-lived. The Model Cities program had struggled to gain the support of local authorities, principally because it distributed decision-making over the whole community. Although this had been the central appeal for Duhl, he later realized that by including so many different groups in the program, "especially if you also put in the groups which included the consumers, you screwed the process up politically from the big-city politician's point of view. As a power base, it wouldn't work for him." As an incredulous Mayor Daley of Chicago had explained to Duhl: "Why should I pay for undermining my own political strength?"[25]

After his initial enthusiasm, Duhl grew evermore disheartened and depressed. In 1968, after less than two years at HUD, he retired from his government role and moved to Berkeley to teach planning and public health. The Nixon administration began to slowly withdraw support from Model Cities, and it was officially disbanded in 1974.

The federal government was shifting its focus back to the rehabilitation of existing housing stock, releasing sixty million dollars in government loans and providing support for "promising approaches to neighborhood preservation which might be adopted by communities on a broader basis."[26] Newman's designs for establishing defensible spaces were embraced as an efficient and comprehensive solution to building new blocks—a means of salvaging existing housing, reducing crime, and increasing property values, at much lower cost to the federal budget. As Newman proudly informed *Time* magazine in 1972, he had been asked by the new Housing Secretary, George Romney, to "prepare a set of specific design directives that HUD could use in all federally-assisted projects."[27]

The alliance that had been built through the Space Cadets—the idea that space was both a physical and a social problem—was now being pulled apart. Despite their initial differences, they had agreed that the better design of physical environments could, when respectful of socioeconomic needs and cultural differences, improve the well-being of citizens. While many environmental psychologists retained their bold and radical views regarding the power of architecture and planning to empower and uplift urban communities, Newman's style of urban design was seen as more efficient and less costly. Because it didn't require the sort of expensive and regionally specific community engagement that Duhl's Model Cities had involved, the Defensible Space design kit could be used in tandem with smaller privatized programs, and it spoke directly to concerns over crime control.[28] Ultimately, although Model Cities might have played an important role in increasing the political participation of minority populations, its housing program was generally seen to have failed.

While Calhoun had imagined that the findings of his rodent experiments would help design better cities, his work was instead increasingly employed to maintain the status quo, used as a rationale for dividing the urban population into separate pens to be controlled more effectively. The criticism of what was becoming seen as a conservative and reactionary architectural determinism is well captured by a particularly damning review of Newman's work in *The Daily Californian* in 1973. Here, William Russell Ellis suggested that while the book might make a good architectural study, as "social science it is terribly weak." Newman claimed to support the ideas of mutual aid and community support, but on closer inspection, he relied on a theory of territoriality that imagined the city filled with aggressive predators, requiring "six-foot cast iron fences and closed-circuit monitoring." If the city was dangerous, it was because of poverty, inequality, and the starvation of public institutions and social programs, not because of design.

Newman's focus on territory allowed the government to prettify public spaces while ignoring the real issues. Ellis caricatured Newman's solution as continuously funnelling residents through a series of defensible spaces, gated communities, and, eventually, to "military encysted suburbs. And then what? An outer layer of welfare cages and humane prisons in some gradient form Pruitt-Igoe to Attica where all 'crime,' 'intruders' and 'hardened targets' retreat to be collected?" As Ellis saw it, Newman was essentially suggesting that the city be designed as a prison or medieval redoubt, in which the problematic and criminal were enclosed and kept separated from the good citizens. Defensible space did not attempt to address the causes of crime, fear, and distrust, and thus, the review concluded, "There is a logic somewhere which accounts for the simultaneous emergence of a Richard Nixon at the apex of his evil and this ambitious proposal which assumes the evil and aims to get it all together 'through design.'"[29]

Calhoun and Duhl's vision of breaking down territorial boundaries between scientific disciplines and working together with designers and administrators to fix the city was being severely tested. For many years to come, Duhl would continue to draw on Calhoun's work to explain some of the problems of urban living, even writing in 1977, "May I ask a crazy favor of you? I am to talk to a group of junior high school kids about the Rats from NIMH. Could you send me pictures, etc. that I can show them?"[30] But in the meantime, Calhoun would return to his laboratory, where he set about designing a new series of rodent enclosures to better communicate his more ambitious and positive solutions to living in an increasingly crowded world.

PART THREE
REVELATIONS

JACK CALHOUN

UNIVERSES, ON AN ORANGE

NIMH, 1960

JACK Calhoun is looking at a picture of an orange, captivated. The orange is partially wrapped in pleated paper so only the uppermost section is visible. The paper's edges are sealed with paraffin wax. Drawn on the exposed cap of peel is a circle in black ink, divided into sixteen radial segments. It looks like the face of a clock. Every other segment is numbered, one through eight. Jack thinks this a marvelous idea.

The enwrapped orange is the handiwork of Carl Huffaker, an entomologist working on the management of agricultural pests in the Experiment Station at Berkeley. To control a species of mite plaguing California's farmers, Huffaker needed to simulate the complexity of the natural environment of the mites in a closed system. But how to miniaturize the world? Huffaker's problem: "There is confusion as to what constitutes a suitable experimental microcosm."[1]

In developing his concept of the umwelt, which would later inform Sommer's idea of personal space, Jakob von Uexküll had imagined the sensory environment of a mite by thinking of only what information it needed to survive. Huffaker has now turned that insight back on itself: What would be the minimal conditions to satisfy the mite's perception that its world was real? He realized he doesn't have to miniaturize the world in which we live, just the world in which the mite lives. Huffaker abstracts. Thinks like a mite, of locating by scent alone the surface of a ripe fruit amid an orchard. He experiments with many arrangements; settles, eventually, on an elegantly formal solution.

He arranges a grid of oranges in a shallow tray, each wrapped in paper sealed with paraffin so only the top section is available, limiting the available surface to enable clean observation. He chooses paper because it keeps the orange fresh; he can maintain an orange for three months like this. In between the oranges, he places rubber balls as pseudo-oranges so he can compare results when the fruit is variously dispersed or concentrated. A wire loop—a walkway for mites—connects the real and the dummy oranges. The clockface dial allows for sampling numbers on each fruit. He introduces the mites. Starting with just four oranges, he builds to a large three-tray grid containing an array of one hundred and twenty oranges and rubber balls. Over many months he meticulously counts the microscopic insects on each clockface, mapping their numbers. Because each tray simulates the world entire, Huffaker calls these abstract grids "universes." He publishes a lengthy essay on his experiments in 1958. In 1960, Calhoun makes a note: he's taken with the terminology.

CHAPTER 16

POOLESVILLE

AFTER the conclusion of the Casey Barn experiments, Calhoun had been awarded a one-year Fellowship at Stanford's Center for Advance Study in the Behavioral Sciences at Palo Alto, California—just across the bay from Huffaker at Berkeley, although they did not meet. Having emerged from his years spent minutely observing his rats from the dark and cramped roof space above the pens of Casey's barn, Palo Alto in the summer of 1962 was space and light and blue skies. A different place, where things were done differently.

Among his colleagues here was John Tukey, a mathematician who had worked with Claude Shannon at Bell Labs in the 1940s. When Shannon had asked what to call the fundamental unit of information, it had been Tukey who suggested *bit*, a contraction of "binary digit."[1] One of the world's leading statisticians, Tukey helped Calhoun to analyze his accumulated data on sequential behaviors in rats, and the rise and fall of their populations. Using Tukey's formulations, Calhoun prepared a paper titled "The Ecology of Aggression" and publishes his monograph-length essay on "The Social Use of Space."

But conventional academic outputs were not the goal. Being among a diverse set of intellectuals at Palo Alto was intellectually stimulating, and almost entirely unconstrained. The sabbatical was a chance to try out new ways of working and thinking. Calhoun spent his days in conversation with the Stanford luminaries, working on his papers. But in the evenings, he had a new project: writing what he called a "post-historic novel," a work of science fiction, set three centuries after the end of our civilization, which he titles *317 P.H.* "P.H." is *post-homo*—after humanity. For Calhoun, fiction became a way of thinking about the future laterally, imagining what might happen if our population crested past the limits of tolerable density and fell into the sort of social stasis he had seen in the late stages of the crowding experiments. *317 P.H.* depicts a world where several hundred million surviving post-humans live in vast reserves in southern Africa. Incapable of creativity, their lives are managed by a small band of scientists who have retained prelapsarian capabilities. Society is organized around family units: "Every Antiquarian City of 1728 persons consists of 144 such families, groups of 12 individuals. Administratively, 12 families form a Horde and 12 Hordes a Tribe, with one representative of each Tribe serving as the Administrative Council."[2] Numerically, it was the same pattern he had tried to implement for the Space Cadets: a way of managing the geometric explosion of possible social contacts, of structuring large numbers of connections while preserving the value of the individual voice within that.

Less prescriptive and certainly less utopian than B. F. Skinner's *Walden II*, for Calhoun, his novel was another way to build a model of the world: *317 P.H.* is also a universe. Conceptually, fiction freed him up, allowed him to pull together threads that ran through his various projects: the induced invasions from the Maine woods, the overlapping home ranges, the thinning of the populations at Casey's barn to avert a population collapse and generate a behavioral sink. As the project progresses, he lapses essayistic, the fictional frame begins to fall away, the novel becomes a notebook. Several of the ideas he would later formalize in academic papers appeared here first. At one point in the novel, each word on a separate line and descending like a staircase across the page, he typed: "The / physics / of the / use of / space / is the / most / important / 'discovery' / in the / struggle / upward / toward—MAN."

ALTHOUGH HIS position at the Center for Advanced Study is a chance for Calhoun to focus purely on theory, he was also thinking of the next round of animal experiments. Before leaving Casey's barn for California, he had been deep in negotiations over acquiring dedicated buildings to continue his research. In 1960, the National Institutes of Health had acquired five hundred acres of farmland on a bend of the Potomac near Poolesville, Maryland, on which to establish the National Institutes of Health Animal Center, or NIHAC. Every possible form of animal experiment ought to be possible here. There was even a six-acre lake, set aside for the study of "mood in waterfowl."[3] Calhoun was excited by the prospect of the site, but out in Palo Alto, he was worried he will miss out on the chance to secure a space for himself.

He sent a flurry of memoranda outlining his research plans, complete with diagrams and blueprints, circulated to colleagues and to his superiors: John Eberhart, Director of Intramural Research, and David Shakow, Chief of the Laboratory of Psychology. It was agreed that NIMH would be allotted an area not less than forty acres for behavioral research, within which Calhoun was promised a facility of approximately nine thousand square feet—roughly the same dimensions as the barn at Rockville. By March 1963, he was told it wouldn't be ready until 1965. By November 1965, that date had been pushed back to 1967.

If there was a reluctance to sign off on Calhoun's schemes, it didn't help that every time NIHAC stalled, his designs became ever-more ambitious. He drew up designs for a series of thirty-seven fully enclosed hexagonal modular units built of insulated metal, eight feet to a side and eight feet high, air-conditioned and hermetically sealed to exclude flies and predators.

The units would be linked with closable walkways to control social interaction, with an observation booth mounted above each. Calhoun called the units of this vast array "space capsules." A network of hexagons, they looked like his schematic representation of distributed home ranges; with their walkways they looked like Huffaker's wire-linked oranges, scaled up for rats. He expected the project would take at least fifteen years to complete.

Eventually, he got a repurposed barn looking down over a sheep meadow and woodland beyond. It was less than he had hoped, but he liked it there, writing to a potential sponsor that the "solitude and surrounding natural beauty helps creativity."[4] It was a two-story galvanized-steel facility, with a large air-conditioned office and twenty rooms on the lower floor for small-animal studies. This was Building 112, initially designated as the Social Observation Study Building, but which Calhoun renamed the Unit for Research in Behavioral Systems, or URBS. He defined a *behavioral system* as:

> a collection of individual animals, living within a definable structured space, whose relationships to each other and to the aspects of the physical space culminate in processes or phenomena not evinced by individuals or not predictable from their behavior as individuals.[5]

The acronym was intended to reference the Latin *urbs*, for city, the etymological root of *urban*, and the ancient Sumerian site of Ur, the first city.

LIKE MANY scientists and academics in the 1960s, Calhoun was deeply influenced by the publication of Thomas Kuhn's *The Structure of Scientific Revolutions* in 1962. Two years prior to Calhoun, Kuhn had also been the recipient of a yearlong sabbatical at the Center for Advanced Study, where he had been struck by how disagreements between the social scientists occurred over much more fundamental questions than were typical among the natural scientists. After watching the sociologists argue, he spent his remaining time at Palo Alto putting together the work that became *The Structure of Scientific Revolutions.*

Kuhn recast the history of science as a series of successive revolutions. The day-to-day business of what Kuhn called "normal science" involved collecting evidence in support of the current theory of how the world worked. If the theoretical framework was faulty, anomalous evidence would begin to accumulate. When the weight of anomalous evidence became too great, faith in that worldview would collapse, and believers would migrate to a new theoretical framework that better fit the data. Kuhn called these worldviews *paradigms*, and the migration to a new worldview was a paradigm shift. *The Structure of Scientific Revolutions* became an unexpected hit, and—much to Kuhn's dismay—the term *paradigm shift* began to be used whenever a popular new theory arrived.[6]

Kuhn had certainly not meant his account of theory-change to be inspirational, but whoever left the house to do "normal" science? It wasn't just the Black Panthers, Parisian students, and Che Guevara who were revolutionary: Kuhn made it possible for scientists to get in on the act. In his notes, Calhoun writes of the possibility of intentionally precipitating "paradigm crises"[7] by generating novel and unexpected experimental results. He wanted to organize his laboratory to maximize the opportunity for serendipitous discoveries, to create anomalies, to upset consensus. "Personally," he writes, "I find the pursuit of puzzles more interesting than the more usual standard normal science endeavors."[8] As read by Calhoun, Kuhn was prescribing revolution.

URBS WAS part of a small cluster of buildings on the south bluff of the Poolesville site set aside for NIMH. The buildings here, backing onto woodland and facing out over the meadows sloping down to the Potomac, became known as the Brain and Behavior Reserve. Calhoun's neighbors included Walter C. Stanley, who had come to NIMH from Maine, where he had been apprenticed under John Paul Scott at Bar Harbor's School for Dogs. Stanley was now working on the effect of human interaction on motivation and learning in neonatal canines—a title carefully worded to conceal the fact that his unit actually played with puppies all day. Stanley didn't much care for the rodent crowding experiments, thinking Calhoun's work stretched credibility, wondering if "there is more than wishful thinking in the hope that your man-rat-man analogy will work."[9]

The main section, under which Calhoun's URBS and Stanley's puppy lab were subsumed, was the Laboratory of Brain Evolution and Behavior, led by Paul D. MacLean. MacLean is a neuroanatomist, now finalizing a grand theory of how the physical structure and developmental history of the brain informed individual and social behaviors. This work involved comparing the neural apparatus of many different animals, and in his labs MacLean had everything: turkeys, squirrel monkeys, even Komodo dragons. He was a brain specialist, and needed good ethologists around to help him with the interpretation of behavior. He quickly struck up a close bond with Jack.

Like Calhoun, when MacLean submitted proposals for his new lab, he aimed big, imagining a central hexagonal facility "with buildings for each of the main subdivisions of the animal kingdom radiating from five to six sides. In other words, there would be individual wings, respectively, for (1) insects and other invertebrates, (2) fishes, (3) amphibians and reptiles, (4) birds, and (5) mammals . . ."[10] What he got was the usual compromise: three long rectangular buildings. But the intended architectural link between physical structure and administrative function was significant. MacLean's life's work had been building up a model of the brain that tied the evolutionary history of the species to the structure of the brain itself.

Back in 1949, while working at the Massachusetts General Hospital, MacLean had published a paper proposing that the human brain was composed of older, more primitive structures, upon which more recently evolved structures had grown. He argued that the "old brain," which he then referred to as the "visceral brain," was connected to emotional functions, while the more intellectual functions were carried out in the "newest and most highly developed part of the brain."[11] After moving to NIMH in 1957, he had identified a deeper layer, a "reptilian" brain, the most basic of all—consisting of the basal ganglia found in birds, reptiles, and mammals.[12]

By the mid-1960s, MacLean's theory was complete. The human brain was effectively composed of three different brains, each on top of the last. At the base was the brain stem, the medulla oblongata, pons, and cerebellum—what he dubbed the *reptilian brain*—which dealt with survival needs, basic motor function, and oversight of automatic processes. Enveloping that was an intermediate zone of neural tissue, which MacLean suggested had evolved to regulate the emotions and the fight-or-flight response. On account of its mediating role, MacLean called this "the limbic system," from the Latin *limbus*, for border. Above this was the neocortex, the distinctive rumpled ribs of the outer brain, which enclosed the brain stem and limbic system. This outer layer, the latest evolutionary development, is more pronounced in smarter animals, most pronounced in humans. This is where the sophisticated cognition happens. MacLean sometimes called it "the thinking cap." The key point was that each affects the other: the reptilian brain is all primal urges, the limbic system tries to control these, and the neocortex seeks to deal with the problems created by being built on top of old technology. As he put it:

> It cannot be overemphasized that these three basic brains show great differences in structure and chemistry. Yet all three must intermesh and function together as a *triune* brain. The wonder is that nature was able to hook them up and establish any kind of communication among them.[13]

MacLean's triune brain hypothesis was an immediate success. It was simple to comprehend, easy to represent visually, and seemed in harmony with so much of our existing folk psychology: the higher functions were literally higher up. Base instincts came from the base of the brain. What Walter Cannon had first identified as the fight-or-flight response from the release of adrenaline and Hans Selye had later linked to the physiological stress response through his General Adaptation Syndrome, MacLean now synthesized into an elegantly simple division of neurological structures. The triune brain also mapped neatly over Sigmund Freud's three-part division of the human psyche into: the primal urges of the *id*; the social self, or *ego*; and the self-conscious reflexive self, or *superego*. MacLean made it possible to salvage much of the intuitive appeal of Freud's account by rebranding its essential elements with modern neurological terminology.

MacLean never claimed the reptilian brain was literally a lizard's brain, nor that other animals did not have the other two layers. His aim was to explain how cognitive sophistication was saddled with the path dependence of our evolutionary history. On account of this legacy, MacLean believed the triune structure had significant consequences for individual mental health and wider social well-being. While humans had immense capacities for reason and emotion, their brains retained primitive urges, which could emerge in individuals and among groups in stressful situations with devastating consequences.

Now working next door to Calhoun, MacLean began to think about what bearing the triune brain might have on the sorts of behavioral pathologies Calhoun had generated in his rats. "Man is not a herd animal," MacLean worried. "Yet the conditions of population increase and city crowdings have in many respects been forcing him more and more into the unnatural condition of the herd during the past century."[14] He began to frame his theory with reference to population growth. Like Calhoun, MacLean feared the rise of social pathology in a crowded world. When he first unveiled his triune brain model to the public in Toronto in 1969, he told his audience: "The most explosive issue, of course, is the problem of controlling man's reptilian intolerance and reptilian struggle for territory, while at the same time finding a means of regulating our soaring population."[15] Robert Ardrey's killer ape was recast in reptilian form.

CONCERNS ABOUT overpopulation had been mounting steadily for at least a decade. In 1957, John F. Kennedy, then nearing the end of his first term as the senator for Massachusetts, gave a speech to the Chicago Economic Club. His topic was the dispensation of foreign aid, but the pretext was "the recent rapid, overwhelming, and utterly unprecedented world population explosion." Kennedy compared humanity to a swarm of lemmings whose "mass death marches" were driven by overpopulation:

> Could this nation, this world, be headed for the fate of the lemming? Could we be plunging blindly on, fat and merry and pugnacious and unstoppable, oblivious to the suicidal course toward which our growing population and economic appetite are leading us? Could this be, not the age of the fatted calf or the golden goose, as we like to believe, but the "age of the lemming?"[16]

In 1960, Austrian American polymath and cybernetician Heinz von Foerster published an essay in *Science* noting how each successive doubling of world population had occurred in half the time of the previous doubling and predicted that if the growth of the human population continued on its current course it would approach infinity on "Friday, 13 November, A.D. 2026." At which point—von Foerster called it "Doomsday"—the need to extract ever-more calories from the biosphere would be redundant: "Our great-great-grandchildren will not starve to death.

They will be squeezed to death."[17] If von Foerster's essay was intended to be playful, it wasn't clear why the punch line was funny. Since the publication of *Our Plundered Planet* in 1948, Fairfield Osborn had been president of the Conservation Foundation, subsequently absorbed into the international World Wildlife Fund. In 1963, Osborn published a follow up, *Our Crowded Planet*, a collection of essays exploring the premise that "the inordinately rapid increase of populations in this world is the most essential problem that faces everybody everywhere."[18]

In March and April of 1966, there were Congressional Hearings on the "Population Crisis."[19] Alaska Senator Ernest Gruening submitted as evidence "Dr. Calhoun's chilling article," "Population Density and Social Pathology," and excerpts from a journal piece called "The Social Dynamics of Population Dynamics." Accompanying Calhoun's writings were Ian McHarg's chapter "Man and Environment" from *The Urban Condition* and Duhl's collection of Space Cadet essays. Len Duhl was called in person as an expert witness, and he pled the case for the environment, stressing that ecologists "have indicated the complex relations between one organism and another, and it is extremely difficult to deal with the population growth of one particular organism without considering its relationship to all others." He called population growth "a problem that is as big as war."

Also called to testify in further hearings later that spring: William Vogt. After the success of *Road To Survival*, Vogt became National Director of the Planned Parenthood Federation of America, formerly the American Birth Control League. He told the Committee: "We are still spending at least 1,000 times as much on death control as on birth control."

The issue reached a crescendo after the publication of Paul R. Ehrlich's *The Population Bomb* in 1968. Although the book initially sold poorly, talk show host Johnny Carson was a fan and invited Ehrlich onto *The Tonight Show*. *The Population Bomb* went on to sell over two million copies. As Ehrlich framed it, the environmental problem was really just the population problem restated: "Too many cars, too many factories, . . . too little water, too much carbon dioxide—all can be traced easily to *too many people*."[20]

The public relations problem with population control was that it was inherently misanthropic. Wishing there were fewer people alive wasn't quite the same as wishing more of them were dead, but the distinction was subtle, and the tub-thumping style of *The Population Bomb* didn't help: "A cancer is an uncontrolled multiplication of cells; the population explosion is an uncontrolled multiplication of people," Ehrlich declared. "We must shift our efforts from treatment of the symptoms to cutting out the cancer. The operation will demand many apparently brutal and heartless decisions." Those heartless decisions included the withholding of foreign aid to countries that failed to implement mandatory sterilization for families that already had three children: "We must be relentless in pushing for population control around the world."

Where during the mid-1960s it had been the dolorous state of public housing that was blamed for crime and civil unrest, focus now increasingly fell not only on the cities but on the planet as a whole. Overpopulation was a global problem. Canadian sociologist Nathan Keyfitz reflects the view from 1969:

> Food riots occur in Bombay, and civil riots in Newark, Memphis, and even Washington, D.C. This ultimate manifestation of population density, which colors the social history of all continents, is a challenge that can no longer be deferred. It will not cease until population control is a fact.[21]

IF OVERPOPULATION did lead to increased unrest and violence, the triune brain offered a novel explanation for why. Here was an additional way to understand what was happening when social strife in the crowded pens led to the breakdown of normal social behaviors. John Christian had concentrated on how prolonged social strife led to physiological consequences; MacLean now supplied Calhoun with a neurological correlate to the "shock disease" Selye had first described. As Calhoun understood this, just as the overtaxed endocrine glands impeded the capabilities of the immune system, leading to elevated blood pressure and eventual death, so, as the higher functions of the mind were disrupted by prolonged stress, were those functions scuffed away, exposing and unleashing the primitive brain below.

As the higher brain controlled the more complex behaviors, it would be these that would be disabled initially. In the case of rats, those higher functions were the social routines—the mating rituals, the deference to dominance hierarchies, and the sanctity of burrows. These behavioral norms, Calhoun had long argued, constituted what could be called the "culture" of the rat. In his crowded pens, culture had been the first thing to go. Unable to endure the endless adrenal alarm bells, the social norms fell away. Calhoun began to think of the social organization of a species—its culture—as a sort of extension to MacLean's triune model. A fourth brain.

With the emergence of increasingly complex human cultures, increasingly dense aggregations of people had formed. Bands became villages, villages became towns, towns became cities. As the limits of tolerable physical proximity were met, our ancestors had, as Calhoun put it, "discovered a new kind of space they could move into." As Calhoun put it: "It was the space of ideas."[22] Crowding—or rather, the capacity to cope with crowding—assumed a central role in the history of human culture. Just as environmental selection pressures drove evolutionary change, so the strife and competition caused by increased population density were key drivers of human ingenuity, as conceptual space was expanded to compensate for the absence of physical space.

As Calhoun followed these ideas, the dividing line between the sort of speculative fiction he had been writing in Palo Alto and his scientific work became increasingly unclear. He charted a series of conceptual revolutions—

the Sapient revolution, the Agricultural revolution, the Religious revolution, the Renaissance, the Enlightenment, the Industrial revolution—each occurring in approximately half the time it had taken to reach the previous leap forward, seeming to trace the curve von Foerster had drawn for successive doublings of world population. Each conceptual revolution enabled an ever-larger number of people, arranged into an ever-larger intercommunicating network.

Why stop there? He projected forward. As population continued to increase, the near future would witness a "Communication-Electronic Revolution" as the computational capacity of the biological cortex to process the required information was exceeded: "This means that we will shortly need electronic protheses which will function much as does our cortex."[23] If Calhoun's calculations are correct, 1988 would see the emergence of "an information exchange network and lead to the development of theories and electronic technologies for the transfer and coding of information as the means for enhanced coping." For dramatic effect, he fudged the date, and in deference to Orwell suggested 1984 A.D.

The runaway pace of population growth wasn't inevitable. If humanity could achieve the sort of electronic-communication network he imagined, he thought it possible to avert von Foerster's Doomsday. The next leap forward could be a "Compassionate Revolution," when our species would unite in common recognition of the need to reduce population levels before they exceeded a critical mass. He posited another possibility: a future of population decline, not increase, ushering in a new phase of human evolution and a blossoming of individual potential. Contra von Foerster, he called that "Dawnsday."

It was all wildly imaginative, and exactly the type of bold and novel hypothesis he had committed himself to pursuing. When von Foerster had predicted that Friday, November 13, 2026 would be Doomsday, professional colleagues knew he was less than serious. What to make of Calhoun's less obviously ludic predictions? MacLean, for one, was apparently willing to take Calhoun at his word. Presenting his concept of the triune brain to an audience in 1969, he opened with a discussion of von Foerster's doubling-curve, accelerating humanity toward infinite biomass. Seeking to offer reassurance, MacLean said: "I harbour some hope for Calhoun's optimistic prediction that just about the time of 'Doomsday' there will in fact be 'Dawnsday.'"[24]

EVEN AS Calhoun mapped out his conceptual trajectory of the future, there was practical work to be done. At URBS, he finally had his own laboratory again. But now, as he began to prepare designs for his next set of experiments, his scope felt wider. At Towson, he had simulated a single city block. At Casey's barn, he had condensed the space while expanding the scale, simulating the distribution of a population in an urban center. At Poolesville, he began to think of the crowded planet. He remembered Huffaker's trays of oranges, how they had stood for an entire world for the mites. Huffaker had called the trays *universes*. Calhoun began to use the same term. From now on, he would build universes.

CHAPTER 17

UNIVERSE 25 / THE KESSLER PHENOMENON

ALTHOUGH Calhoun will be undertaking a new set of rodent experiments, he doesn't intend to go back over old ground. By the mid-1960s, Calhoun's crowding experiments are being replicated in different settings and using different animals. He is frequently consulted on their design. Confident of his conclusions, rather than repeat his experiments he urges other scientists to diversify, alter initial conditions, play with variables. When psychologist Harold D. Fishbein, working on how children conceptualize space, writes from the University of Cincinnati, Calhoun advises: "I do not believe that exact replication is the essential necessity. What would be desirable would be to have some critical tests of the general hypothesis underlying the notion of a behavioral sink."[1] In 1963, he is asked to advise on a study by Alexander Kessler, a physician interested in whether population density affects genetic inheritance. Kessler is taking a hiatus from his career as a medical doctor to undertake a Ph.D. at the Rockefeller University in Manhattan, where he plans to study how crowding affects the heritability of fur color in mice. Calhoun sends him some notes with detailed designs for a large octagonal enclosure, ten feet across, that should allow him to breed a small population of mice at elevated social densities over several generations, with advice on thinning the population to avert a premature population collapse.

Meanwhile, at Poolesville, Calhoun is busy planning his own series of experiments. In Building 112, he replicates the observational setup he had first designed for Casey's barn: on the ground floor is a suite of large, sealed rooms, with glass panels cut into the ceiling to permit total observation with minimal disturbance to the animals. At Casey's barn, he had four such rooms. At Poolesville, he has eighteen, each capable of housing anything from ten to one hundred rats. To investigate the precise effects of environment on social behavior, he plans to test separate variables in each room—density, space, diet, lighting, and so on. The operation was to run for several years. He envisages monitoring activity using an automated computer system.

Each room will be considered as a closed world, what he now calls a "universe." He works up a formal definition of the term:

> An experimental universe is a bounded physical space consisting of one or more similarly constructed cells. Each cell provides opportunity of the expression of many of the behaviors characteristic of the species for which the universe is intended as a place of habitation.[2]

Where the enclosure at Casey's barn had four such cells, the position of the ramps meant that the end cells provided a defensible space. This caused the population to aggregate in the central pens and led to the development of a behavioral sink. In his new universes, each cell will be identical, and equidistant from every other.[3] There will be no impassable barriers between the boundaries of adjoining cells, only low dividers that would break up the floor like garden fences, allowing for more privacy and helping to generate a sense of shared space among different groups. They were screens, not walls: suburban picket fences. The more formal design of the universes represented a compromise between the type of single-variable studies being conducted by experimenters such as Joe Brady at Walter Reed and the open-plan compound studies.

Calhoun's broad goal is a more precise account of the circumstances that lead to the development of social and behavioral pathologies, and the creation of strategies to mitigate the negative effects of crowding. Where previously he had been happy to allow others to extrapolate from his rodent studies to human societies, he will now explicitly design the universes to model whole cities, perhaps whole societies. When he first thought of cramped city blocks in the mid-1940s, the problem seemed peculiar to the urban core. By 1967, the population explosion has made the issue global. Commensurate with his new ideas about triggering a conceptual revolution, Calhoun is thinking big. The universes of URBS are models of human civilization in total. His premises are larger, his conceptual ambition vastly more so.

Characteristically, he overreaches, realizes he has neither the staff nor the technology to begin the experiments as designed. The computerized system of monitoring rats he has in mind awaits advances in microprocessor development, while NIMH have the budget for only a handful of employees. He gets seven: five lab staff, plus a secretary and a computer programmer. When he advertises the vacancies, he warns prospective candidates that they should not expect a bustling workplace environment: "We are for the present exposed to a certain amount of physical isolation and the incumbent in the present position will have to accommodate to these circumstances."[4] With his earlier plans for the sci-fi "space capsule" enclosures rejected, and his rodent universe studies unfeasible, he is stalled. Laboratory space is at a premium, and Calhoun is acutely aware there are plenty of other researchers who would happily occupy Building 112 if it stands idle. He needs to rethink his plans, and fast. His problem is that he doesn't have a puzzle.

And then he hears back from Alex Kessler, whose multigenerational study on the heritability of coat color in mice has run into a somewhat unexpected situation. As Calhoun had been an advisor on the project, Kessler invites him to New York to see what has happened:

> He took me over to show me his two populations of mice. Each was contained in a tub-like octagonal metal container only five feet across. One contained over 800 mice and the other over 1,000. No mouse could move without being in contact with one or more others. The speculative possibility of "standing room only" had been achieved!

This was astonishing. The pen was teeming with mice, more than eighty per square foot. They were as a solid mass, a fluid agglomeration of seething fur. But there was not the rabid violence and reproductive failure Calhoun had always seen in his own experiments. No behavioral sink. The mice were at standing room only, and they were still breeding. According to standard population ecology, this should be impossible.

Something in the design of Kessler's environment had completely disrupted the usual phases of population growth. Populations—whether bacteria, fruit flies, rodents, or humans—tended to follow an S-shaped growth curve: increasing slowly as the first few colonizers adapted to their habitat, then growing rapidly as breeding took off before entering a phase of territorial and social competition as density increased, which increased rates of morbidity and mortality while dampening reproduction, causing the curve to flatten off. This "terminal growth phase" should have resulted in severely stressed rodents, but the opposite seemed true. "Looking at Kessler's mice I got quite a different impression," Calhoun remarked. "Behaviorally speaking, they appeared washed out."[5] The mice were almost serenely passive, seemingly indifferent to the constant press of other bodies around them. Although such behavior as was observable seemed severely attenuated, reproduction continued—partly as a consequence of the sheer density. Pups were scattered haphazardly among the throng of adults: "Enough of them survived despite little maternal care just because often enough a lactating female would stop momentarily on her trips to no where in particular."[6] And hence, "whenever a lactating female stopped, she was likely to be on top of a youngster needing milk."[7] It appeared that Kessler had, quite accidentally, achieved unheard of levels of population density without the concomitant violence and associated behavioral pathologies. Calhoun had found his puzzle. He returned to Poolesville intent on reverse engineering what would become known as "the Kessler Phenomenon."[8]

Checking the notes and plans he had originally sent Kessler against the setup he saw in New York, Calhoun discovered Kessler had deviated from the recommendations in two crucial respects: he had used a smaller enclosure, and he had used a larger seed population. Calhoun had suggested an enclosure ten feet across. But at the Rockefeller University in the heart of Manhattan, Kessler had found it difficult to secure the required space, so he had scaled the design down to a little less than half that size. Meanwhile, when Calhoun seeded a new enclosure with what he had once called "the first colonists," he deliberately chose to use only a very small colonizing population. The intention was to simulate the manner in which animals arriving in a new habitat would come to occupy that territory, which in the wild would likely involve only a few individuals: "That is the way uninhabited places get most often get populated—some wandering pregnant female or wandering pair accidentally come upon such paradises. Ecologists 'know' that this is the way things happen." In Towson, Calhoun had used four pairs for a quarter-acre enclosure. In Casey's barn, just one mother per cell. But Kessler wasn't interested in simulating the arrival of a species into a new habitat; he wanted to test the effects of crowding on genetic inheritance, and he wanted to get the generations up and running quickly. And so, rather than the recommended one or two pairs of mice, Kessler had begun his study with sixteen pairs, in a pen with an internal diameter of only forty-eight inches, what Calhoun called "oversize wash-tub like boxes only an arms reach across."[9]

The central mystery was the passivity of the crowded mice. Behaviorally speaking, there ought to have been no significant difference between rats and mice in response to social density. And so, partly to follow Kessler's study, and partly for convenience, Calhoun made a compromise: "We decided to turn to mice, since their space requirements are less than for rats and cheaper to construct. Space, budget, and the demand to be 'up and at it' imposed restrictions on what ideas might be pursued."[10] The mice he chose were an albino strain called BALB/c—a variety that had once been curated by Clarence Little at Bar Harbor, and of which NIHAC had plentiful stocks. The white fur was easily marked with colored dyes to identify individuals as the experiments progress, plus the strain was physiologically stable, robust, and easy to use in the laboratory. Kessler—who was looking at the inheritance of fur coloration—had been using four different strains to form a founder colony that would yield visibly different pelts across generations. But four different strains created an unpredictable hybridity not only in coat color, but also in behavior. Calhoun wanted to control for behavior, and as all BALB/c mice were genetically alike, as similar as identical twins, by using the single variety he could be assured that any observed behavioral differences "must solely arise from inequality of experience."

With his URBS team, Calhoun now set out to systematically test the relation of enclosure size to colonizing population size. Unlike Kessler, he had plenty of space, so he planned to run a series of experiments concurrently, systematically measuring the relationship between the size of the colonizing population and the size of the enclosure. He designed five different universes, scaling up in size: a single-cell enclosure, a two cell, a four cell, an eight cell, and a sixteen cell. The sixteen-cell universes were so big they wouldn't fit in the ground floor observation rooms, so he planned to have them on the upper floor. He wanted to build five of each type, in which he would test five different sizes of colonizing population: one pair, two pairs, four pairs, eight pairs, and—like Kessler—sixteen pairs. That made twenty-five enclosures in total. Twenty-five universes.

But as they began constructing the enclosures, he realized he had again been too ambitious. A quick calculation revealed that if anything close to Kessler's numbers were reached, after only a few generations they would be dealing with a population of thirty to forty thousand mice—far too many for the limited staff he had. Calhoun gambled: he abandoned the colonizing-population variable and decided instead to run the experiments with an identical colonizing population of eight mice in each of the five sizes of universe.

By mid-1968, construction of the five universes was finished, and each seeded with four pairs of young mice. As expected, when the mice reached sexual maturity breeding in the initial phase was rapid, but even with this reduced plan, the populations very quickly became unwieldy. After only a few generations, the five extant universes of Building 112 housed over five thousand mice. This presented significant problems. "Our little group of investigators included only eight persons, including the secretary and computer programmer, both of whom were frequently drafted into making observations and recording data," but even with the goodwill of his colleagues, Calhoun conceded "the task was becoming impossible."

A year into what was looking like an expensive misstep, URBS cut their losses. They would learn nothing more from the variable sized enclosures. But Calhoun felt the original intent of the experiment had been realized: given the breeding rates they were witnessing in each of the five universes, it was clearly the size of the colonizing population, rather than the size of the enclosure, that was affecting the population growth and the tolerance of crowding for the subsequent generations. Calhoun now understood Kessler's "error." Initiating the study with so many colonizers in the same space had made it impossible for the mice to establish their own territories. In the absence of territorial domains, mice growing up in the already-crowded enclosure had never developed any concept of personal space. Without any personal space bubble, there was a much greater tolerance of proximity. As Calhoun put it: "Ego boundaries stop with the individual's skin. Only a direct contact with sufficient force to penetrate the hide could induce a stress state." With this new lifestyle of high-density living and lowered rates of social stress, more offspring would survive, but they would do so by sacrificing their sense of self-identity: "This was the theory I set out to examine as a possible explanation of the puzzle formed by the anomalous terminal high densities observed by Kessler."

With their resources stretched, the URBS team made a decision to terminate the populations in all but one of the enclosures, leaving only the largest and most elaborate of the structures: Universe 25. This larger sixteen-cell environment would provide them with the best opportunity for studying complex processes over time, and they "decided to follow it to the bitter end, or through an unlikely recovery through renewed breeding." The experiment was to become the most well-known of all the work Calhoun ever conducted. It would also be the last experimental cycle he ever completed. In it, he would observe the entire history of a population from first colonization to eventual extinction. Universe 25 was a closed space into which no new immigrants would enter, from which no individual could leave. "Whatever might happen within the universe," he declared, "the mice had to stay and face it."

UNIVERSE 25 was a square pen built on the upper floor of Building 112. It was almost nine feet on each side, enclosed by four-and-a-half-foot-high galvanized metal walls supported by an exterior wooden frame. This box was open from above, with a sheer face around the lip to prevent escape.

Inside, the floor area covered a little over seventy square feet, but the walls were lined with wire mesh to allow mice to climb up to the water and feeding areas above. Behind the mesh, tunnels led to the nesting spaces. He had designed these as "one-room apartments," stacked vertically. Two hundred and fifty-six apartments in all. The universe was divided by low barriers into sixteen cells, radial segments fanning out from the center, lined with a substrate of ground corncob. Each wedge-shaped cell contained four sets of apartments, four water bottles, and a food hopper. Two mice could comfortably drink simultaneously from each bottle, fully twenty-five mice could simultaneously eat from each hopper. An old can filled with nesting material was set on a repurposed lab stand in each cell.

What Calhoun called Phase A began when the four pairs of immature young colonists were introduced on July 9, 1968. With space abundant, they distributed themselves to the four corners of the enclosure and established territories. The first litters were born after one hundred and four days. After that point, a phase of rapid exponential growth began, with the population doubling approximately every fifty-five days. This was Phase B. As the population grew, more and more of the universe was occupied. When the population crested over six hundred, every cell was full. Like the colonies in Towson, the reproductive rate followed a dominance gradient: the most fecund cell produced one hundred and eleven pups; the least, only fourteen pups. At this point, there were over three times as many immature mice as socially established older mice, but so far all had received a good upbringing: "Motherhood flowered, most young survived, and every mouse received an early education fitting them for mouse society."[11]

On day three hundred and fifteen, a second, slower period of growth began. Calhoun notated this as Phase C. The population now took five months to double its size. With no predation or disease, the quantity of elderly mice began to accumulate. Many senescent individuals could no longer breed, but still held social status, and did not vacate their properties.

With the remaining apartments already occupied by established territorial males, younger males were automatically of low status simply because all the available territories were taken. Only very rarely did a young male successfully challenge an older male. Trapped within the metal walls, the juveniles found themselves with nowhere to establish their own domain, and no opportunity to emigrate.

More and more young males now gathered on the floor of the pen, doing less and less. Calhoun noted how "the frustrated, rebuffed younger males begin to avoid their older associates, and participation in courtship activities with females declines, then disappears."[12] Rejected first from possession of property, and latterly excluded from reproductive activity, they began to huddle in listless pools at the center of the pen where the divider spindles converged to form a sort of no-man's-land. In Calhoun's terms, they had "ceased trying to acquire a normal social role." Although these mice were still physically present, and relatively healthy, they played no part in the social life of the universe: "Socially speaking, they had removed themselves from the universe, they had psychologically emigrated."

Every few hours, the silence of the huddle would be broken by sudden spasm of violence. These episodes were caused when males returning from feeding would attempt to bore down into the huddle of withdrawn mice, seeking shelter. This sudden disturbance to the group frequently triggered panic in one of its members, causing it to "go berserk." The panicked mouse would attack any adjacent mice. Although the resultant wounding was often severe, none of the victims ever fled from or resisted these attacks. Calhoun was amazed by this passivity: "Even the flight response, which we had thought to be quite instinctual, had been washed out."

Meanwhile, the young females retreated to the upper stories, occupying any vacancies in the highest and therefore least-favored apartments, where they could expect some level of protection within a stable colony. They were not safe here for long. With an increasingly large population of single males now active on the floor of the pen, territorial males defending their familial nest boxes faced a near-constant barrage of challenges. Exhausted, many of these once-successful fathers simply gave up their positions, abandoning the apartments and the females and young within. Now unprotected, the nursing mothers were consequently subjected to frequent nest invasions. Brutalized, they became violent in response. This aggression was displaced into the home, as females began to attack their own offspring—treating them as if they were interlopers. The wounded pups would be driven out of their nests before they were old enough to leave, emerging improperly socialized into the arena below. Calhoun describes the hopeless future faced by these rejected youngsters:

> They started independent life without having developed adequate affective bonds. Then as they moved out into an already dense population many attempts to engage in social interaction were mechanically disrupted by passage of other mice.[13]

As a result of the nest raids, pregnancies did still occur, but fewer pups survived to weaning. Later, autopsies would reveal that many embryos had been reabsorbed. Sometimes, the mothers would try to relocate their pups, but the young were frequently abandoned. Litters that had been counted by the team had often simply disappeared by the next survey. The period of population growth was coming to an end. "For all practical purposes," Calhoun grimly notes, "there had been a death of societal organization by the end of Phase C."

Maximum population density occurred at day six hundred. There were now over thirty mice per square foot of floor area. For comparison, the rat rooms at Casey's barn had a floor area of approximately one hundred and forty square feet, and a population that never exceeded ninety rats. Universe 25 had a floor area of just seventy square feet, and, at its peak, some two thousand two hundred mice. The final phase now began. Calhoun called it Phase D, or "the Death Phase."

IT IS here that the environment of Universe 25 diverged from his previous studies. Where the compulsive "pathological togetherness" drove the rats to increased violence in Casey's barn, the mice of Universe 25 had slipped past stress. Rather than destroy one another, these animals were seemingly content to simply abide. The mechanism allowing this involved a sort of developmental retardation.

Although few young were born, the almost five hundred mice that matured in this final phase grew up in a social desert where almost no meaningful interaction took place. The offspring of maternal neglect, driven out before they could be properly socialized, they had never developed a sense of personal space, had no desire to assert or defend themselves, nor any urge to claim the territory of others. Calhoun called them "a type of organism never before encountered":

> Males, even when fully adult, essentially never became aggressive, nor did they court or copulate. They had become asexual and asocial. Not fighting, no scar tissue marred their pelage. Instead, they spent an excessive amount of time grooming—to them there was not much else to do other than to eat, drink and sleep.[14]

These were the Beautiful Ones. Not only were they tolerant of permanent proximity, they seemed to require it. Males began to perch on the metal rings of the lab stands where the paper bedding had been stored, each in contact with another, but always facing apart. They had never outgrown the craving young pups have for being pressed together: "Males as adults accentuate within-litter seeking of body contact."[15]

Consistent with his understanding of MacLean's triune-brain hypothesis,

Calhoun believed they had never developed the ability to perform any of the complex behaviors their species was capable of. Physically mature, the Beautiful Ones were stalled in early infancy. He compared them to thirty-five-year-old adults with the minds of an eighteen-month-old baby. The usual range of communicative activities—vocalizations such as squeaks, physical gestures such as rearing up or boxing one another—had ceased altogether. A strange silence and stillness characterized the pools of mice. They were physically together, but socially separated. On account of the severely limited behavioral repertoire and absence of mutual engagement, Calhoun described such males as "autistic."

No less strange were the females. Where the males gathered in silent flocks at the center of the pen, the females roamed the walls and higher surfaces. Almost all these "rejected females moved up in the highrise apartments."[16] They became entirely fearless and uninhibited. The normal shyness and caution around new objects was replaced with an obsessive curiosity—they would gather around any new object placed within their enclosure. When Calhoun stepped into the pen to perform cleaning duties—refilling food hoppers, removing waste and dead mice, replacing the substrate—rather than scurry and hide, the females would instead move toward him, clustering around his feet in swarms of up to a hundred. "If I began walking slowly about the floor," he observed, "they would form a murine wave trailing behind as I circled round and round."[17] Inverting the usual metaphor, Calhoun dubbed these mice "pied pipers." Developmentally he estimated they had not proceeded further than forty-five-day-old mice, the equivalent of a human at around five years of age.

With population densities so high, it seemed inconceivable they were not stressed, and forensic analyses were ordered to check for physiological markers. When Calhoun had been running the Casey barn project, he had sent his samples to John Christian. At NIMH, Calhoun referred his histological analysis to his friend Julius Axelrod, who was the world's foremost expert on the activity of adrenaline—work for which in 1970 he was awarded the Nobel Prize for Physiology or Medicine. As expected, in the autopsies of the males from earlier generations who had been subjected to violent attacks during the initial phase of crowding, Axelrod found elevated traces of catecholamines, indicating the animals had lived in a state of near-constant severe stress. But autopsies of the late-stage Beautiful Ones revealed no such traces; their adrenal metabolism was indistinguishable from the socially successful territorial males. Or, as Calhoun later described it: "Both types of male were relatively unstressed, both the dominant males who could cope, and the autistic beautiful ones which could detect no reason to cope."[18] While it was of course impossible to know their true psychological state, as far as their bodies had recorded the somatic evidence, theirs had been happy and successful lives. Despite their utter failure to engage in any meaningful social activity, they had apparently existed in a state of docile contentment.

Crucially, the population didn't rapidly die off, nor descend into destructive violence. Calhoun's universe was beginning to imitate Kessler's. He had reproduced at a larger scale the psychologically dislocated population he first saw swarming in the washtubs at the Rockefeller University. The mice of Universe 25 had found a stable strategy for coping with extreme population density. They were at standing room only, but entirely relaxed about that. Sedentary, calm, content, and physiologically healthy, they continued to live like this for a further two years. But what was good for the individual was catastrophic for the species.

A cyclical boom-and-bust pattern is entirely normal in many natural populations. It is present in the historical cases of swarming lemmings and mouse plagues that had been recorded by Charles Elton, and it had been the fate of Christian's deer on James Island whose shrinking landmass compressed the herds into intolerable levels of proximity. In the wild, when a population died back, the survivors rekindled their numbers to a more sustainable quantity. But here, the ability to restart society had been lost. Prior to reproductive senescence—the rodent equivalent of menopause—a sample of late-phase female mice were removed from Universe 25 and relocated to one of the vacant smaller universes that had been constructed for the abandoned parallel studies. Here they were introduced to healthy males from fresh breeding stock, but while some litters were born, the young were not properly raised, and none survived past weaning. The mice had lost the ability to reproduce.

Meanwhile, among the remnants of Universe 25 where no social interaction occurred, breeding rates declined toward zero. The last successful conception was recorded in mid-January 1970, nine hundred and twenty days into the experiment. By March 1972, the average age of survivors was a little over two years of age—a life span at the upper limits of a mouse in ideal circumstances, significantly longer than the age range expected in the wild, where they rarely exceed six months. During the almost two years of the protracted "Death Phase," the mice gradually died from old age. By mid-July 1972, the population had fallen to fewer than a hundred survivors. The utopian state of peaceful coexistence at more than double the optimum density had been achieved, and its consequence was the slow termination of the population by social stagnation. As Calhoun put it, "all destined to die as Methuselah."

The final male died on December 5, 1972. Adjusted to human years, he was one hundred and two years old. The last female died a month later. The URBS team had come to know her as the "Grand Old Lady." On January 8, 1973, one of his researchers found her on the floor amid the corncob and paper, cold and listless. Calhoun had taken her out of the now-depopulated universe, placed her under a bulb for warmth, and tried to encourage her to feed and drink. "I watched her all day," he said, "and at one point I thought she was going to make it but she didn't." In human years, she was one hundred and eight.

He later said there was no mourning because this generation had no personality, no self. "They weren't mice," he explained. "They had motion, but it had no meaning, it's the minimum motion needed to keep the machine going." The corpse was packaged for autopsy, the soiled universe dismantled. The Grand Old Lady was the last mouse, but she had died long before her body stopped. Jack shrugged: "You can't identify with nothing."[19]

CHAPTER 18

POPULARITY CONTROL

WORD of the experiments spreads long before they are completed. Bob Cook, the founder of the Population Reference Bureau, had watched Calhoun's ascent since helping him secure employment at NIMH back in 1955. In 1970, Cook urges his friend, the *Newsweek* columnist Stewart Alsop, to visit Calhoun at Poolesville. Alsop agrees, arrives at NIHAC on a sunny July afternoon, enters Building 112, and later emerges equal parts appalled and amazed. "The experiments are fascinating, all right," he grumbles, "but they spoiled a beautiful day for me." *Newsweek* publishes a full-page editorial in August 1970—almost three years before the termination of the project, and long before significant results had been collected or published. "Dr Calhoun's Horrible Mousery" had no photographs, but Alsop describes the experiments in detail, and vividly evokes "the rank mouse smell, which is at first overpowering."

When Alsop encourages Calhoun to make links between his socially disengaged mice and the youth of contemporary America—"Aren't we maybe seeing the phenomenon of the Beautiful Ones, already, in the drop-out drug culture?"—Calhoun demurs, saying that as a scientist he couldn't possibly speculate. But, with off-the-record prompting, Alsop is happy to do that for him:

> This generation of the young, unlike their elders, will live to see Dr. Calhoun's "upper threshold" reached. Is it possible that when the threshold is reached, population growth will be ended, not by birth control or the bomb, but by the mysterious and terrible process that ended all reproduction in Dr. Calhoun's horrible mousery? Is it possible that the young have some sort of subconscious prescience of what lies in store?[1]

Alsop's piece is to set the tone for the reporting of Calhoun's work throughout the 1970s, while *Newsweek*'s broad circulation brought the activities of URBS to an international audience. Calhoun recalls it as a turning point: "It set off a deluge of newspaper and television coverage that gave wide visibility to our findings long before we had published any of the details in more customary scientific publications."[2]

They come from all over the country, thrilled by the strangeness of Universe 25. *National Geographic* dispatch journalist Robert L. Conly to Poolesville. *The Rocky Mountain News* in Denver, Colorado, sends a correspondent to report on how "Crowding Produces Weirdo Rodents," which the journalist describes as "an animal counterpart to the 'Bowery bum.'"[3] The *Smithsonian Magazine* profile "The Small Satanic Worlds of John B. Calhoun" calls his Poolesville laboratory "a pet shop in hell." Pointing to the socially etiolated young mice of the third phase, the article emphasizes how the ongoing damage will be intergenerational:

> Dr. Calhoun suspects that they can never again be made whole. He extrapolates that belief into a baleful suggestion

> that has not occurred to most of our social scientists: Taking a child from a slum environment and giving him some sort of ideal life is never going to repair the damage done in his earliest years.

On the back of this, the *Smithsonian* calls him a "prophet."[4]

That theme of crowding being especially damaging to the young also features in an interview to *The Kansas City Times*, where Calhoun is, again, apparently reluctant "to draw a parallel between these mice and the drop-outs of today's youth culture but insists that the breakdown in socialization that occurs in overpopulated mice is transferrable to human beings." Only a year after Woodstock, *The Kansas City Times* sees in Universe 25 troubling indicators that, for the new generation, the gainful pursuit of productive and creative endeavors may be replaced "by participating in mass gatherings or by taking drugs."[5] The locus of concern is shifting from violence to ennui.

In 1971, *The Washington Post* runs the first of what will be several articles on Universe 25. Although the subheading declares "Colony Overpopulates to Extinction," at the time this claim remains unsubstantiated—the experiment still has two more years to run. The *Post* writer, Tom Huth, delights in the spectacle of the pied-piper females compulsively following Calhoun's feet as he steps down into the floor of the enclosure, and the scene is caught by NIMH photographer Nilo Olin. The bizarre images of Calhoun stood in the spoked well of the pen with a flock of mice trailing him circulate widely. They draw the attention of broadcast media. He repeats the stunt for the cameras when visited by a CBS film crew, and again when visited by McGraw-Hill Broadcasting, a TV network recently acquired from Time-Life: Muscle Jack, scaler of chimneys, now in his mid-fifties, nimbly hops over the wall into the universe, where he gently scoops up dazed mice to display for close-ups. His work is subsequently detailed in NIMH's *Progress Reports* with a celebratory six-thousand-word account of Calhoun's career to date and plans for the future.[6]

One of his favorite articles is from the short-lived but influential business magazine, *Innovation*, which runs an in-depth profile of the URBS lab in 1971. In "It's Not Every Day You Walk into a Laboratory Whose Mission Is to Save the World . . . ," journalist Nina Laserson takes considerable time to fill in the backstory of Calhoun's research, and—more importantly for him—is willing to explain his radical forecasts for humanity's future. In part, this indulgence is a consequence of *Innovation*'s unusual business model: subscription only, ad free, aimed at the entrepreneur, each issue of the magazine is designed by a Madison Avenue agency with bespoke illustrations and handsome layouts. They even commission a painting in which the French revolutionaries of Delacroix's *Liberty Leading the People* are replaced with mice—an image of which Calhoun is immensely fond. With its editorial mix of novel ideas and practical suggestions, *Innovation* is a precursor to TED Talks and management books, and a commentary at the end of the article helpfully includes suggestions for their corporate audience: "Shouldn't you have at least one person in your organization who is the kind of environmentalist thinker that Calhoun calls for?"[7]

Geared toward the inspirational, the *Innovation* article is one of the few to focus on Calhoun's optimistic belief that humanity could successfully escape the fate of his overpopulated docile rodents by engineering a conceptual revolution in the coming decades. Unfortunately, this positive message of creative disruption reaches no further than the exclusive circle of *Innovation*'s well-heeled subscribers; with an annual membership of $75, it is later described as "one of the most expensive magazines on the planet." [8] More typically, the reportage of Calhoun's rodent studies is set somewhere between macabre horror and quirky comedy. When his last mouse, the Grand Old Lady, dies, Tom Huth of *The Washington Post* marks the occasion with a mock obituary notice:

MOUSE, THE LAST

On. Jan 8., 1973, at the Laboratory of Brain Evolution and Behavior; of terminal overcrowding; no survivors; no services; no matter.[9]

IN HIS willingness to entertain the press, and to partake in their sensationalism, Calhoun had been playing a dangerous game. Journalists were useful as ciphers to make connections between his rodent experiments and human populations, but he was unable to control the editorial slant, and the reports did not always bring him the audience he sought. As the press attention ramped up, Calhoun found himself courted by the increasingly prominent population control lobby. In 1969, he had been invited to a conference organized by Fairfield Osborn's Conservation Foundation, which was described as "A Conversation on Population, Environment, and Human Well-Being." Also in attendance: Humphry Osmond, Ned Hall, and *Population Bomb* author Paul Ehrlich. Calhoun used the occasion to promote his positive vision of humanity's potential for escaping the psychological damage caused by crowding through a concerted expansion of conceptual space. The minutes record that "John Calhoun dissociated himself from the attitude of pessimism which he felt pervaded our meeting and emphasized our duty to try to 'participate in a more rapid process of cultural change, without which our visions will die.'"[10] At this point, the agendas of population control, environmentalism, and mental health seem reasonably well aligned. Over the coming years, those issues would markedly diverge.

In 1971, Calhoun attended a seminar at the United Nations in Geneva to discuss potential themes for a conference on world population planned for the following year in Stockholm. Among the representatives was Alex Kessler, accidental architect of the first "standing room only" environment, and now inaugural director of the Human Reproduction Programme of the

World Health Organization. Calhoun returned to America disappointed that the focus was narrowing in on contraception, preventing future people being born rather than optimizing conditions for the already living. His work had long been cited in promotional material arguing for the importance of domestic birth control—mainly among the urban poor[11]—but since he had seemingly plotted a new form of demographic catastrophe, he became attractive to organizations seeking to limit global population growth. They assumed him a natural ally.

He was not. Their rhetoric and motives worried him. In 1969, *Medical World News* spoke of "people pollution," suggesting that psychological disorders due to crowding could soon be more deadly than physical diseases, and that "as the decade draws to a close, Dr. Calhoun finds that a growing number of investigators all over the world agree with him that his rat and mouse studies hold clues for mankind's survival."[12] He began to receive letters from supporters of the ZPG, an activist movement seeking "Zero Population Growth," cofounded by Paul Ehrlich.

The ZPG were admirably clear about their aims: "The ultimate goal must be a zero population growth," declared Walter Howard, an ecologist who specialized in vertebrate pest control; "the birth rate must not continue to exceed the death rate."[13] Many were willing to consider "involuntary fertility control"[14] to achieve this. Even when Calhoun's name was not explicitly invoked, the equation between social density and behavioral derangements that he had done so much to promote was frequently employed in support of the ZPG's goal. "Crowding unquestionably nurtures man's baser instincts and predisposes him not only to violence and criminal behavior but to a callous inhumaneness," declared Harvard Medical School's Professor of Population Studies, Roy Greep. Consistent with the logic of economic inflation rates, Greep held that a surplus of humans degraded the value of their individual lives: "In India, starving beggars lie on the streets near death, but the throngs pass by or step over them with total lack of concern. Life cheapens in its own abundance." Or, as Walter Howard put it: "The destiny of overpopulation is the erosion of civilised life."

Calhoun felt exasperated. Ever since reading Eugene Rochow's chemist's-eye view of "maximum human protoplasm" back in the late 1940s, he had been frustrated that the population problem was being framed as either a resource-scarcity problem or an access-to-contraceptives problem. As Calhoun saw it, "though they are important issues, they are not the essential issues, and are certainly not the ones of most importance in the long run."[15] Writing to his superior John Eberhart, who oversaw the unit of which both he and MacLean were a part, he complained: "So far other agencies have given primary attention to such issues as birth control and food production in regard to population crises. Similarly with regard to the environmental crisis most attention has been given to pollution and safeguarding the biosphere."[16] He urged that NIMH should capitalize on their "unique capacity" for addressing the social and psychological issues relating to social density. Overpopulation was also a mental health problem.

Meanwhile, Calhoun's particular objection to the ZPG wasn't simply the coercive nature of their proposed means, nor even the skewed focus on reducing growth primarily in developing nations, but that the goal itself was faulty. "I believe that the ZPG doctrine is a very insidious one," Calhoun explained, "from the evolutionary point of view."[17] Zero population growth implied a world of total demographic stagnation: a settled, static scattering of isolated tribes into which no immigration, out of which no emigration, and between which no competition. If it was a world that seemed at peace, it was also the world of docile, huddled mice he was looking down at over the rim of his universe.

When he worked with Davis and Christian in the Baltimore row houses, social strife had been the crucial component that first spurred his thinking about population regulation. As he later learned in Casey's barn, at high levels, this social strife created unrest and violence. But he now knew that the total absence of social stresses—the sort of environment he was presiding over during these final years of Universe 25—was, in the long term, no less damaging. Although the individuals were physically healthy, without at least *some* social pressure there was no social interaction and no adaptation; and with no capacity to adapt, the species died out. Even in Towson, he had detected that the emergence of a dominance gradient among the rat colonies had been essential for generational advance, writing back then that "if rats failed to develop a class structured society they would not long survive as a species."[18] Universe 25 was the experiment that confirmed that early supposition. To make his point, he compared the aims of ZPG to the lung fish in Lake Manyara, Tanzania:

> For millions of years they have maintained a constant way of life avoiding crises. Their major crisis is the drying up of lakes. They have learned to hide and not to cope. As the lakes dry up, each lung fish burrows down into the mud, forms a mud ball cocoon, reduces its metabolism, sleeps out the trying, drying period. We too could devise our solitary cells of intellectual isolation to wait out each and every crisis.

"I for one," Calhoun added, "would not like to be a lung fish in Lake Manyara in Tanzania."[19] A stable population represented an "evolutionary dead end."

REMARKABLY, NEARLY all of this attention and popular acclaim preceded his first scientific report on Universe 25, which was made to the Royal Society of Medicine in England in late summer of 1972. This was his chance to correct any misinterpretations of his studies that might have arisen from the press reports, and to explain how this latest round of rodent experiments was more than a repetition of his earlier work. Things did not go to plan.

Calhoun had been invited to present at a symposium on "Man in His Place," held at the Linnean Society of London, just off Piccadilly's Burlington Arcade. Here, he would be speaking in the same wood-paneled theatre in which the joint announcement of *The Theory of Evolution by Natural Selection* by Charles Darwin and Alfred Russel Wallace had first been made. He was thrilled to have the chance to speak from the same lectern: "Behind it in 1858 the theory of evolution made its debut!"[20]

Perhaps emboldened by the recent press adulation, or simply excited by the venue, Calhoun decided to swing big. He called his paper "Death Squared: The Explosive Growth and Demise of a Mouse Population." Although the main portion of this lecture was a procedural description of the findings of his Universe 25 study, he bookended the sober scientific material with his more radical ideas. If—in his discussions with MacLean about "the fourth brain," and his private writings since Palo Alto speculating on the future of mankind—Calhoun had been veering toward the quasi-mystical, he now gave full flight to his prognostications. He began his lecture by declaring "I shall largely speak of mice, but my thoughts are on man, on healing, on life and its evolution."[21] By the fourth sentence, he was quoting from the Book of Revelations, invoking the Four Horsemen of the Apocalypse and the Tree of Life, and speaking of "the death of the spirit." This spiritual death, following the biblical account, was the "First Death," and bodily death was the "Second Death." At the end of his introduction, he suggested: "Perhaps we might do well to reflect upon another of John's transcriptions (Rev. ii.11): 'He who conquers shall not be hurt by the second death.'"

He was making an entirely reasonable distinction between the physiological death of the organism and the psychological withdrawal he had documented in his rodent populations. But his phrasing and scriptural references tilted his presentation toward the mystical.

When the lecture ends and the questions begin, Calhoun realized he had badly misread the room. His audience ignored almost entirely the bolder ideas, and burrow down on details: What about pollution in the universe—how frequently was the bedding and substrate changed? What about the colored dyes used to mark individuals, might this affect the perception of other mice? Most upsetting for Calhoun: when he brought up the issue of "conceptual space" in humans and mice, he was chastised by the Chairman, John Zachary Young. The eminent Professor Young was the Establishment personified: ten years' Calhoun's senior, educated at Marlborough School and Magdalen College, Oxford, he now chided Calhoun for comparing rodents to humans. As the minutes record it: "*The Chairman* interrupted Calhoun and said that he ought to be careful about this. He had been discussing mice." Calhoun recalled the same: "Before completing my remarks I was interrupted by the chairman of the session, Professor J. Z. Young, the famous comparative anatomist. He warned me rather strongly about making extrapolations from mice to men. I was left nearly speechless."[22]

Calhoun stepped out of Burlington House hurt and embarrassed. "I left that session markedly depressed," he later wrote, "and could not face joining the group for dinner that evening." Instead, he paced the streets of London, returning to his hotel to ruminate on why the talk had seemed to go down so badly. In part, his failing had been rhetorical: the conceptual and the scientific material were mismatched. Not only was the juxtaposition of precise methodological description and broad metaphysics tonally jarring, but the two aspects—experiment and implications—seemed to be of wildly different orders of conceptual magnitude. An audience excited by the philosophy was likely to be turned off by the experimental detail, while anyone sufficiently knowledgeable about rodent behavior was likely to be puzzled by the inclusion of the philosophy.

Meanwhile, in the balance of content, he had perhaps made the metaphysics seem an optional extra, real science with some philosophy on the side. But it wasn't detachable. The metaphysics was inextricable from the experiment. And this in turn because the behavior of the rodents—their decline into complete asociality—was entirely a product of the environmental conditions in which the rats were housed. The disappearance of complex behaviors, what Calhoun characterized as the contraction of their conceptual space, wasn't an evanescent phenomenon, floating free of the physiology, but an effect every bit as real as the wound scarring or enlarged adrenal glands. Although he had perhaps erred in using biblical quotations to convey his meaning, the science was sound.

The apparent failure of the audience to grasp what he was trying to say, and his failure to properly communicate it, was frustrating. But most of all he was stunned by the rebuke from Chairman J. Z. Young. If Young didn't think there were valid lessons to be drawn about human society, what did he think Calhoun was doing? Universe 25 was almost entirely pointless if that inference was disallowed, merely a pit of multicolored mice, closely observed from birth to slow death. For Calhoun, whose entire scientific career had been spent on the working assumption that the mental activity and behavior of rodents were essentially continuous with the mental activity and behavior of humans—no less comparable than the functioning of the lungs or the endocrine system—to be accused of naive anthropomorphism was deeply upsetting. And to face that accusation in the very room where the unveiling of Darwin's theory had collapsed the historic distinction between humans and the animals!

BEING SO publicly challenged on the relevance of his work to human societies clarified something else for him, something he had said at the very start of his lecture: "I shall largely speak of mice, but my thoughts are on man."

Back in 1962, Ned Hall had written to John Christian to encourage him to seek a wider audience: "Why not start working on human populations applying the same techniques that you used in your mammalian studies,"

Hall suggested, "what about examining the adrenals of criminals and people who die in violent deaths in over-crowded slum conditions"?[23] Christian had declined. Not necessarily because he wasn't interested in the consequences for humans, but because that wasn't his field of expertise. He was a population ecologist specializing in physiology. Christian's message was internal to the professional community with whom he already communicated. Do the deer of James Island encode a message about human civilization? Perhaps, but that sort of connection was not his to make; Hall was an anthropologist, let him do that. John Christian had kept his focus on the animals. Jack Calhoun had gone the other way.

Once upon a time, Calhoun had been an animal experimentalist whose work might have an indirect bearing on the lives of people. But between the Towson enclosure and Universe 25, gradually but inexorably, that priority had reversed. The amelioration of human life in a crowded world had become his primary mission, and rodents had become his tools. Besides, it was by now too late to retreat to the sort of conventional "normal science" that his former colleagues were pursuing. In the building just next door to his laboratory, MacLean had sought to revolutionize the study of the human brain, launched to broad acclaim the sort of daring hypothesis that aimed to change the way an entire field considered its subject matter. MacLean had taken the risk and appeared to have pulled it off. Calhoun gambled on being able to do the same.

CHAPTER 19

A PRESCRIPTION FOR EVOLUTION

WHILE the press was interested in his work, Calhoun had reciprocated, collecting newspaper clippings on gang violence, rioting, rapes, murders, a collage of reports on international conflicts, environmental crises, sexual deviance. He marks up novels and philosophical treatises whenever he sees parallels with his rodents. The emergence of new psychopathologies interests him. There are notes on what is now called "battered child syndrome," first described in 1962 as a "clinical condition in young children who have received serious physical abuse, generally from a parent or foster parent."[1] He thinks of how his crowded rats and mice had attacked their own offspring, wonders if he has found an animal model of the process by creating "mouse parents" that "ceased loving their offspring."[2] Calhoun thinks of the social isolation of his Beautiful Ones.

As the violence of the 1960s and early '70s peaks, he expects to see greater levels of social disengagement. He notes the rising case reports in the psychiatric literature and popular press of autism and emotional withdrawal among children. Within ten to fifteen years, will the new generation be born into a world so crowded that—like the mice in Universe 25—they will never fully emerge from an infantile state?

All feels urgent. Calhoun believes humanity needs to consciously organize its own future as a species. He has traced a history of population growth and conceptual innovation moving together in lockstep, a series of cognitive revolutions that have permitted humanity to tolerate elevated social density by escaping into "conceptual space." As this process could not continue indefinitely without reaching the crisis point von Foerster called "Doomsday," and as Calhoun believes that a static population as envisaged by the ZPG would result in a social and cognitive stagnation, he is confident that the population will soon begin to decline. But he is also conscious that as the population decreased, the quality of life of the average individual would be threatened. A declining and ageing population would require the younger population to take on a wider range of social roles, while requiring more advanced technologies of data processing and communication to avoid being "lost in a sea of information."[3] He imagines that coping with this declining population will be a spur to innovation, ensuring that humanity will continue to adapt and evolve in the place of earlier advances impelled by increasing population density.

He decides he must confront the issue head on, make explicit what he had meant when he spoke about "the death of the spirit" and the prospect of "Dawnsday" when humanity would undergo a "compassionate revolution." He needs time to think.

IN EARLY 1974, Jack writes to MacLean and John Eberhart to request a sabbatical. His previous sabbatical had been the year at Palo Alto, twelve years earlier. But where his stay at the Center for Advanced Study had thrust him into a busy group of freewheeling thinkers, this time he plans to study and think alone. Intellectual progress involves a division of academic labor between specialists and generalists, the details people and the big-ideas people. Diligent researchers like John Christian collected and analyzed experimental data within their field, while synthesists and popularizers like Ned Hall were able to rove widely, collecting it all together into overarching theoretical frames. Calhoun is now trying to do both roles himself.

Compounding that, he is a prodigious notetaker. The quantities of data he has collected have always outstripped his capacity for analysis. The Towson experiment concluded in 1949 but remained largely in note form until the publication of *The Ecology and Sociology of the Norway Rat* in 1963. And there is still material from that period he feels needs more analysis. He also wants to revisit the work from Casey's barn in the early 1960s, plus some of the studies he did in Bar Harbor, including all the "induced invasion" fieldwork conducted by his Census of Small Mammals network. As he explained to MacLean and Eberhart: "Effective synthesis of such data requires a total immersion in it for long periods of time without distraction."[4] His request is approved, and he is granted six months private study leave, beginning June first.

Each night during the last week of May, he arrives back at his home in Bethesda from Poolesville with his car full of papers, books, and reels of computer tapes and printouts onto which his numerical data has been transferred. Holed up in his basement office, surrounded by his materials, he is ready to begin. He has made it clear he is to remain undisturbed: "I informed everyone that I would answer no phone calls (only calls answered could be from Edith by time or code arrangement)."[5]

To record his progress, he decides to keep a sabbatical logbook. In the first entry on June 4, 1974, he lays out his plan:

> It will take me some time to develop a strategy for most effective utilization of my time. At first I will work back and forth through the material to become refamiliarized

> with the whole context. This process will in part continue in order that common themes or problems across studies can be interrelated.

The next entry does not appear until September 10:

> As is apparent from the lack of diary postings, I have been remiss in following the course of my sabbatical.[6]

IT SEEMS he began earnestly enough, but the more he rereads his earlier notes, the more he felt there was to add. Rather than resolve into a linear and progressive narrative, each trawl through his archive turns up more leads for new projects. He begins to feel overwhelmed by all the undone work: the accumulated drift of unfinished essays, incomplete experiments, unrealized plans.

There are other motives for retrospection. Carol Houck Smith, a senior editor and later vice president at publisher W.W. Norton, contacted him a few years earlier. After reading Stewart Alsop's *Newsweek* piece on "Dr. Calhoun's Horrible Mousery," she wrote to say how she was "unable to forget his account of your work and its possible ramifications." She thought other readers would feel the same, asking if he could write a book on "mice but mostly men?"[7] He has also been in discussions with several university presses interested in publishing an academic monograph. The attention is flattering, but the requirement to assess his own significance weighs upon him.

He then gets distracted when Edith, who was working at the National Library of Medicine, indexing medical terminology, brings home a box of Xeroxes containing citation data. Jack begins poking through it, a diversion at first. Mired in the detail of his own research, he realizes that here is another way of thinking about information. The librarian views content differently to the scientist. Where Jack is tied up with the meaning and significance of his work, Edith views the information not in terms of its connections with the world, but in terms of its connections to other information. One is not simpler than the other, neither is more true. But here is another way to process information: not to evaluate, only to connect.

During that summer and fall, he does begin writing a book: filling several hundred pages of typescript with descriptions of his research, interspersed with autobiography, speculative plans and designs for experiments, philosophical musings, and even some poems and drawings. He titles the manuscript $R_x$evolution” to the symbol used by pharmacists for denoting a prescription, R_x. Dated November 14, 1974, it remains unfinished, dissolving into clippings and handwritten marginal notes. He later confides to a friend: “Edith thinks I will never write a book since I always get diverted to something more pressing after I am half way through and the ideas generated in the process set me off into a somewhat different path.”[8] And so it will be. For by the end, he has a vision of where he needs to go next.

He persuades Eberhart and MacLean to extend his sabbatical, which was due to end on December 31, 1974, to June 1975. He will come into the office at Poolesville for two days a week, spend the rest of his time working on the plans for his next project. As he will later tell a journalist from the *Baltimore Sun Magazine*, “I was brooding in my basement office at home, looking over my past”:

> And I went through a sort of conversion experience, seeing what I should be doing in the next eight years before I retire. I asked myself how I could make the greatest scientific contribution, one that had the highest probability of application to the issues relating to population increase. I wrote it down and the administration bought it.[9]

What the administration bought is an ambitious cycle of three thematically linked and overlapping studies. Having assessed his life's progress, he calculates it takes him approximately ten years to plan, execute, and analyze a crowded rodent habitat, and a further two to four years for publishing and circulating his results.[10] In 1975, Jack Calhoun is fifty-eight years old, and conscious that he is running out of time. This will be his last cycle of research, his last chance to do something revolutionary.

THE FIRST of these projects is effectively a repeat of Universe 25, but this time, octagonal rather than square, and almost twice the size—fully seventeen feet across. Mice in the square enclosure had clustered first into the corners, leading to uneven initial density. So he shaves the corners off, making each of the eight interior sides equally desirable. This new pen is also radially segmented, but into only eight cells. Along the walls, these cells are split by a low divider, forming a perimeter of sixteen subcompartments. Unlike Universe 25, the radial spokes dividing the cells do not meet in the middle; the central forum is a communal zone shared by all. Therefore, the further a mouse goes from its nest box, the more opportunities for social interaction are available. This will be Universe 33.

Unlike during his previous experiments, he knows exactly what will happen. When the population reaches around eight times the optimal density: "No mouse will ever again engage in reproduction, the population will slowly age, and death will take its toll until the last surviving senescent subject dies." Then: "Extinction!"

So why do it again? Because the first time around, his URBS team did not anticipate reproductive behavior among the mice would cease quite so early, and consequently failed to properly observe the steps leading up to the emergence of an asexual population. In Universe 33, each cell has more internal dividers, which should encourage a greater number of subgroups to form, which will hopefully act to retard the onset of what he now calls "the ultimate pathology." The aim is to attend more closely to how behavioral complexity evaporates.

THE SECOND project is new. If the first aims to generate the "ultimate pathology," the second aims to demonstrate how humanity has so far managed to avoid it. Per his reading of MacLean's triune-brain hypothesis, Calhoun believes that prolonged stress-related crowding causes the limbic system to be overwhelmed. Depending on the intensity and duration of such stresses, the complex functions of the neocortex—MacLean's third brain—are disabled. This results in either the release of untrammeled violence and sexual behavior (the behavioral sink) or withdrawal into asociality and asexuality (social autism). In the former case, the ability to perform complex behavior is lost; in the latter, it never develops. Calhoun's addendum: Humanity has developed an ability to endure far greater levels of social contact by escaping from limited physical space into a progressively expanded conceptual space. His question: Could rats learn to do the same?

And so with this second project, he wants to help rats to cope with crowding by enlarging their conceptual space. Obviously, he doesn't expect his rodents to become cultured in the human sense. But he does think that he can encourage more of them to demonstrate more complex social interaction. "Our objective," he explains, "is simply to encourage them to acquire sufficient culture, compatible to their small brains, to offset the pathologies produced by an eight-fold increase in density above the optimum." If he can show that more cooperative rats are able to tolerate higher levels of population density, he will have some degree of experimental evidence to back up his theories about how better communication has helped humanity face the same.

For this experiment, he is using rats because they are smarter than mice, and generally capable of a greater range of more complex behaviors. He opts for a hybrid variety, a combination of white Osborne-Mendel lab rats and wild-caught stock. The brown rats will be sourced from Parson's Island—native home of the first Towson colonists—and hence a population he already knows intimately. Recalling Richter's discovery that wild rats are more intelligent and much less docile, and in response to recent criticisms from a growing number of psychologists that the inbred albino lab rat was "degenerate"[11] and an "idiot,"[12] Calhoun wants to maximize the scope for cognitive development. Ideally, he would have used only the Parson's Island rats, but there's a pragmatic consideration: "It takes between five and ten times as much effort to conduct the same type of study on experimental sociology with wild rats as it does with domesticated ones," he writes to a fellow researcher. "At this stage in the development of ideas it would have been impossible to obtain the much larger funds required to conduct comparable studies with wild rats."[13] So really, by interbreeding, he's not making smarter lab rats so much as slightly dumber wild rats.

One of the ways he intends to gauge conceptual development is by measuring what he calls the "social velocity" of his rats—the frequency of their interactions. High-velocity rats are socially promiscuous, more dominant, and usually more successful at reproducing. Low-velocity rats are more solitary, more submissive, less likely to breed. But where the extroverted high-velocity rats will lead predictable and normal rat lives, it's among these low-velocity introverts that Calhoun sees promise: "Low-velocity individuals tend to withdraw from social interaction, but if not too withdrawn, they tend to be more creative; they develop innovative patterns of behavior."[14]

He calls this style of enclosure a *velocity pen*. The layout is very different from his other studies. Many years earlier, when short on indoor laboratory space at Bar Harbor, he had once built vertical mouse apartments. He resurrects something like that design now, but much larger: tall rectangular boxes—his original designs are twelve feet high—with three flights of "vertical ramps" mounted up the walls. These vertical ramps—like drainpipes with internal ledges—lead to three tiers of upper-floor balconies running around the inner walls of the enclosure. The three tiers of balconies provide access to the nesting apartments. The whole space is split into two symmetrical sections, which mirror one another, each nine by nine foot, forming a total floor area of one hundred and sixty-two square feet. The drawings look like postwar concrete housing projects, balconies facing in on deep courtyards. He plans to construct two such pens, designating these identical structures as Universes 34 and 35.

Between every compartment, every balcony and stairwell, food area and drinking area, an electronic portal is fitted. The rats are to be implanted with miniature glass lozenges containing a metal coil. When a rat passes between compartments, an electromagnetic field logs the movement. Every movement of every individual, twenty-four hours a day, will be logged by computer. "This technology has an aura of Orwellian Big Brotherism," Calhoun admits. "We believe our use of it is a highly positive one." That "positive" use involves teaching the rats cooperative social roles. As the population grows, he will allow the population in one of these universes to simply do its own thing. This is the control group, Universe 34. To the other enclosure—Universe 35—he will provide tasks. At each doubling of the population, a new task will be introduced.

This is social engineering for rats. He wants to make them more cooperative. To do so, he has designed a "socialization training apparatus for obtaining water," or STAW. There are two compartments, separated by a plastic partition. For any rat to get water, there needs to be another rat in the apparatus. This will allow each rat to press a lever and obtain water. As soon as one rat moves away, both levers will lock, and thus the rats will learn to drink in the presence of others. He believes that rats conditioned to behave cooperatively will have learned a new concept: altruism. He calls the setup a "culture-inducing universe" and believes it will make the colonies more stable and capable of coping with density.

He wants to incubate a sense of "tribal affinity."[15] Because rats favor a nocturnal environment, he decides to paint each half of the symmetrical universes a different color: a light daytime and a dark nighttime. This will help the rats develop a clearer dominance gradient, with the high-velocity rats occupying the preferred dark side and the low-velocity introverts on the light side. This will be further compounded by his second task, organized around eating: an electronic gate will permit either the high- or low-velocity rats to eat in the food resource compartments only when those of their own "clans" are present. If any representatives of both tribes or clans are present, a door automatically drops over the surface of the food hopper, blocking access to food. He thinks that if he can create a greater number of discrete social groupings, the population will divide into smaller and more stable networks that will help to reduce the likelihood of unwanted interactions. This will also mean more social roles among the population. Where the inhabitants of Universe 25 were largely "socially unemployed," in Universe 35, all the rats will have something to do. A purpose.

New residential areas will be made available as the population expands, and density will be carefully regulated by removing surplus pups. As the population level is slowly elevated, he is confident in his hypothesis that the "cultured" rats of Universe 35 will develop larger and more complex communication networks and be able to endure greater levels of social density than their counterparts, left to organize themselves, in Universe 34.

As he explains in an article for *Populi*, the journal of the United Nations Fund for Population Activities: "Just as our forebears shaped our culture for us, so we as researchers will provide culture for our rats."[16] When interviewed by the *Baltimore Sun*, he is more audacious, explaining that—if his process pays off—they may expect a conceptual advance in the rat that "takes it forward to about the beginning of higher primates, not quite up to the chimp, if we're correct we expect to make as big a change in our rats in six years as nature with evolution and natural selection did in 50 million."[17] He is planning to make "super rats."[18]

The premise is as ambitious as the setup is technically challenging. Even with the computerized monitoring system, it will require a much more interventionist approach than anything he has done at this scale. But although it looks superficially like a brutalist high-rise, there's something else going on, too. When he produces a flowchart of how the various spaces, chambers, and rooms are connected, it reveals "the environment to be bilaterally symmetrical with some connections between the left and right sides."[19] Basic operations such as food and water at the lowest level, an intermediate set of floors for normal social activity, and a top floor available for the more complex tasks. The whole thing bilaterally symmetrical and linked with connecting tunnels. Life-supporting operations at the base, higher functions at the top: the entire enclosure is also a physical representation of MacLean's triune brain.

ALTHOUGH HE has designed what looks like high-rise housing with Orwellian levels of social engineering and constant monitoring, he certainly isn't suggesting anything like Universe 35 should be scaled up for people to live in. This is a model for rats, and built for their particular needs. And so his next question: What would it look like to build a system for expanding the conceptual capacity of humans?

His third project is not an experiment in the usual sense, but it is certainly experimental. Communication is the key. He had been looking at his rodent communities as communication networks that transferred information between individuals and groups, some more successfully than others. In Universe 25, as the society began to break down and the animals became more "autistic-like," they lost the capacity for social communication; even the vocalization of the mice essentially ceased. That had been both the means by which they tolerated the social density and the ultimate cause of their extinction.

Sat in his basement office during the first months of his sabbatical, surrounded by the carloads of material he has brought back from his office, he has felt overwhelmed by the mass of data, and by the missed opportunities that seem to unspool from every project. And he is only one researcher. How much more information is lost like this? The current system of intellectual communication—journal articles, monographs, chapters, and conference papers—is overburdened and inefficient. Important findings are often ignored, buried among the stacks, only to be independently rediscovered years or decades later. As the global population increases, the problem of accessing information will only become worse even as the need to do so becomes more urgent. Some form of enhanced network of communication among experts must be created.[20] And so the third of his projects will not involve building a rodent universe, but building a communication network, initiating what he describes as "a frontal attack on development of the prosthetic brain."[21] What he is planning is, in essence, an attempt to model a way in which human conceptual space might be expanded.

Looking back, the idea of a communication system has been there all along. It was the means by which the colonies in Towson distributed themselves across the quarter-acre pen. It was the mechanism by which voles and shrews kept apart in the Maine woods. It was there in the loose network of rodent ecologists he enlisted to compile the North American Census of Small Mammals, and with his Space Cadets, whose rolling cast of twelve experts drawn from different organizations, disciplines, and systems met twice per year to share data, methods, technology, and ideas. Since the mid-1960s, he has also collected over three thousand excerpts from academic papers on environment, population, and mental health. The corpus of text is unwieldy, the information it contains is no more accessible for having been placed in one location. Realizing now that "information retrieval required indexing," he begins to group the excerpts by assembling strings of keywords that overlap with themes in other pieces. He begins to build a "conceptual network of related terms."[22] It is a librarian's way of thinking about science.

As is so often the case with his projects, the problem turns out to be bigger than he anticipated. Edith's box of citation data has shown him how the author might be thought of as the "center of an invisible college consisting of those that cite his work."[23] He now realizes that each concept can be thought of as existing at the center of a similar web of related meanings. Choosing the proper level of taxonomic specificity proves fiendishly difficult:

> Single excerpts were found to contain at least two, and sometimes as many as five terms that essentially conveyed the same implication. Furthermore, different authors would use different words when dealing with essentially the same idea. If my author's-own-term index was to prove useful, a means had to be developed to pool similar terms to represent generic concepts.[24]

As he curates his index, the keywords he uses begin to detach from academic disciplines, locating more abstract and primary concepts: not psychology, biomedicine, physics—but violence, tradition, creativity. These begin to form oppositional poles: *aggression* opposes *compassion*, *hope* opposes *pessimism*. But binary axes don't allow enough connections. He plays around with the available links. Tries grids, tries trees. Neither provide enough vectors. He arranges concepts in a radial dial, like the surface of Huffaker's orange, like the spoked floors of his universes. When he mapped the interlocking home ranges of adjacent species in the Maine woods, he schematized these areas as tessellating hexagons of different sizes, interlocked like a surface of Islamic tiling. He thinks now of the shrews and the red-backed voles, how they lived at different heights in the same space. He reads a 1964 report from the RAND Corporation about the likelihood of discovering habitable planets that plots the distribution of potential sites as equidistant in space. Inspired, he folds his diagram up, makes a three-dimensional structure. The triangular segments of the dial become the triangular faces of a polygon. Their vertices are twelve nodes, each touching twelve others. The shape becomes a multifaceted crystal, a gem. It is an icosahedron, a twenty-faced shape with twelve corners.

The structure of the icosahedron allows him to organize his conceptual nodes in such a way that they have a sufficiently complex but still manageable level of connections to the concepts: “It means that every generic concept imbedded in the network will be linked to 12 other generic concepts. Similarly, concepts located at the periphery of the network will be linked to six other concepts.” Twelve groups of twelve: it is the ideal size of the rat populations in a city block, it is also the number of connections he has long believed is ideal for optimal human relationships.

THIS, THEN, is the third and final project: not a rodent universe, but a universe of ideas. It isn’t yet the “electronic communication network” he thinks will become necessary in the near future, but a preparatory attempt to model how best to organize knowledge so as to allow for the maximum efficiency of information exchange.

Jack Calhoun anticipates crises in the coming years, and he is trying to build a network to enable rapid exchange of information between experts when those crises arise. He calls it a “Global Alerting System.” This is not an easy sell. But it is this that he has built Universe 33 to advertise. And, when his sabbatical ends in June 1975, it’s this triad of projects that he presents as his research proposal to carry him from now until retirement. In essence, the first project—his octagonal mouse habitat, Universe 33—demonstrates the erosion of the “third brain” and the loss of complex behavior due to crowding. The second project—comprising the “culture-inducing” Universes 34 and 35—deals with expanding the conceptual capacity of rats to enable them to cope with increased crowding. The third project is a human experiment: an attempt to model a "fourth brain," a way of organizing information that will permit greater global communication.

Diagnosing the various pathologies created by crowding is no longer his central focus. The population will continue to increase, this is a given. What no one else is offering is a way of dealing with that. And so he is trying to design something else: not a rat city, but a strategy for coping with being around so many people in human cities. A solution that doesn't call for involuntary sterilization or the conditional deprivation of foreign aid. This is what he calls his "prescription for evolution," a next step,

> in which we will foster the evolution and interconnectivity of machines that process information. They will gain a life of their own, a life augmenting our own as we should be augmenting the life of all the less cognizant species on earth. At some far distant point in the future, perhaps a hundred thousand years from now, the worth of all these machines may also become a constant. Right now I can't worry about the next crisis in evolution, the time when our appendages, the information processing machines like ourselves now, also get boxed in.[25]

Calhoun's perception of the rodent universes begins to alter his perception of the world around him. Perhaps this has always happened. Perhaps he first began to see the world as a closed universe when he built the pens at Casey's barn, or perhaps Towson. Or perhaps even before then, in the back alleys of the Baltimore row houses. He begins to perceive the commonalities flowing in both directions now:

> You and I live in such relatively closed universes of interconnected cells. We live in communities of

> apartments or houses arranged in blocks. We work in office buildings with roads, halls and doors connecting various sized cells. We go to stores or restaurants, each of which is a cell connected to similar ones by a network of roads. How we develop acquaintenceships depends in part on how often we see someone, and how often we use them in part depends upon how the interconnectivity of the cells of our environment influence the likelihood of us being in the same cell at the same time.

His rodent universes, built to simulate the world about him, now overlap and occlude the place they represent. Our world, too, is a universe of cells. "I raise this issue now," he adds, "to get you to start asking the question: 'How much more intelligent is my behavior than that of the rat with respect to being blown about or trapped by such vagaries or winds of chance?'"[26]

CHAPTER 20

SYSTEM FAILURE

SINCE its founding in 1949, NIMH had grown to become the largest of the National Institutes of Health, at one point accounting for over a fifth of the total NIH budget. In 1967, it budded off to become its own division of the Public Health Service, equivalent to a bureau and equal in status to the NIH itself. In 1974, NIMH was subsumed into the Alcohol, Drug Abuse, and Mental Health Administration. Under ADAMHA stewardship, the priority was pursuing only such research as could be directly applied in practice—preferably toward drug and alcohol issues. As an ever-greater share of the budget was funneled away from basic research and into treatment centers, former NIMH psychologists complained that "research is shortchanged when it is administered by an agency that also administers services programs."[1]

WHEN CALHOUN returned from his sabbatical, it was with renewed enthusiasm and purpose. He immediately set about trying to convince his superiors of the importance of implementing

> a communication network designed (a) to channel scientific and philosophical expertise to consideration of the course to and the nature of possible future states of society, and (b) to facilitate the value reorientation these states of society will demand by both the general public and by policy and action oriented agencies.[2]

It was an idea he had first developed a decade earlier but kept confined to memoranda and personal correspondence. Cognizant of the stage of his life, inspired by the publicity surrounding his previous research, he was ready to begin his revolution.

What he was now calling his "Alerting System" was a distributed and interlinked network of experts, capable of detecting imminent crises arising from world population, and of marshalling the resources to respond. Although he envisaged a global network of scientists and other experts, each—like his rat colonies, like his Space Cadets—in local teams of twelve, he thought in the short-term NIMH might lead the way. He approached Director Bertram Brown, suggesting they initiate a small-scale version of his Alerting System using staff at the Institute. He wrote letters to leading scientists, statesmen, politicians. In 1974, he even writes to President Nixon, inviting him to spend some time at the URBS lab to discuss the future of humanity. Facing the coming "megacrisis" of soaring global population, he warned the president that the world now stood at an important junction of history, at which civilization faced a choice between "extinction, mere survival, or into a vision spreading a bright aura over past accomplishments. Your decisions, by default or intent, will make the difference." He helpfully added that the Poolesville site had "Adequate helicopter space."[3]

In the early days of NIMH, when Calhoun and Len Duhl had cooked up the idea for the Space Cadets over lunch, there might have been tolerance for this type of thinking. Now, it was at odds with the restructured and more practical priorities of ADAMHA-NIMH. What were the measurable outputs? Where were the impacts? A standoff awaited. Calhoun had long considered himself an outsider to the establishment. He ran the operation at Building 112 like a commune. At URBS, no titles were used, and he shared office space with his staff. One of the journalists who had visited recalled her surprise at finding the senior investigator "in coveralls in the middle of a huge mouse pen, doing the dirtiest of dirty work."[4] In part, though, he did the mucking-out duties because no one else was around to do it.

There was no Eugene Casey this time around, and external funding had been more difficult to source. The Menil Foundation, based in Houston, Texas, had initially pledged an annual donation of one hundred thousand dollars for six years, but that deal had fallen through. The new administration of ADAMHA-NIMH grudgingly agreed to continue support but could not provide him with the fifteen staff he requested, and would need to evaluate his outputs. In 1976, he had ten employees. By late 1977, there were only five. As the workload for each remaining member increased, stress-related absence among the URBS team rose in step. What he called his "personnel crisis" triggered a personal crisis, as he feared the whole project would collapse. He frequently complained about the difficulties of "working within a government bureaucracy, with all its traditionalism, rules and constraints."[5] Jack Calhoun was feeling the strain, and things were about to get worse.

ON A bright fall day in late October 1978, the NIMH Board of Counsellors visited Poolesville to review the work being carried out in the Laboratory of Brain Evolution and Behavior. The morning was scheduled for presentations, the afternoon set aside for a tour of Calhoun's URBS rodent experiments. His octagonal mouse habitat, Universe 33, had been up and running since 1976 and was proceeding much as expected. The scene was calm. Having passed through the initial phase of violence and competition, the mice were now settling into their terminal coping mechanism of total withdrawal. "Once a population reaches this stage, there is no way to undo the damage," he explained. "They are bound for extinction."[6]

The other rodent experiments were advancing more slowly. After a series of delays, he had finally begun the culture-inducing habitat, Universe 35, only a few weeks earlier, in September 1978. Consequently, the demonstration of his facility was less impressive than he might have hoped, but the computerized portals were installed and working, and, as the first cohort of sixteen rats moved around the universe, each individual's precise location was recorded. In time, he would accumulate over twenty-five million separate data points. But to the Board of Counsellors in October 1978, he had little more to show than a high-tech rat cage and the theories contained in his initial proposal.

An article in NIMH's internal newsletter had announced the arrival of the first colonizers of Universe 35 with the headline "Rats Given 'Culture' As Defense Against Overcrowding," going on to explain how:

> Overcrowding takes its toll on a species' most complex behaviors, according to Calhoun. In the case of rodents, these have to do with territoriality, attachment to place, caring for others, the young, courtship and mating.
>
> Calhoun theorizes that the 'social networking' and cooperative role playing learned in the programmed environment help the animals resist the breakdown of such complex behavior.

On Calhoun's copy of the article, he would later write: "This proved correct." Much else he hoped to learn from the study would never be completed. What he called a "glitch (unresolvable)" forced them to abandon plans for a male-female pair-bonding aspect of the experiment, requiring that opposite sexes come to the water source to be able to drink.

Still, by spring of 1979, he was beginning to see some significant results. In the control group, Universe 34, where the rats were left to conduct their own social affairs, there was notably more violence and social division. The dominant rats even plugged the portals with paper to prevent the lower-status rats from entering their half of the enclosure. Meanwhile, among the cultured group of Universe 35, a more harmonious community emerged. Later that same year, he believed he was seeing enough behavioral changes to warrant calling them "super rats." "In our observation of rats over the past 30 years," Calhoun reported, "nothing closely comparable to the behavioral diversity and rapidity of adaptation expressed by these 'super rats' has been observed."

All proceeded well. By 1980, he was witnessing a reduced need for "status interactions" among the cultured rats, and—more surprisingly—a reduction in reproductive rate among females. The wariness of human contact—usually absent in docile lab rats, but bred back in from the wild strain—had been retained among the control group, but among the cultured group, they not only tolerated being handled but seemed actively curious, spending more time in the open and "peering at investigators entering the habitat. Neither wild nor domesticated strains have previously been noted to behave in this manner."[7]

WHEN IT came back in late 1978, the report from the NIMH Board of Counsellors had not been good. They had criticized the breadth of his approach, and suggested he focus on only one of the three projects. MacLean offered what support he could, but nevertheless, Calhoun would be forced to shorten the length of his studies. Reluctantly, the decision was taken in 1980 to terminate the crowded mouse experiment in Universe 33, its remaining population euthanized long before they could die of old age. The study wasn't entirely without merit: the more complex and spacious design of the enclosure had indeed extended the period of "behavioral and social integrity" far longer than had been seen in the smaller and simpler Universe 25. And although he would have liked to see the experiment through to its natural conclusion, Calhoun felt they had already gathered adequate information on the various steps by which "a complex living system develops and then disintegrates in the absence of opportunities for non-genetic evolution."[8]

More disappointingly, his cultured-rat study in Universe 35 ended in 1982—much earlier than he would have liked. He had hoped to observe his rats through four successive doublings of population, envisaging the emergence of "neighborhood groups" that integrated high- and low-velocity rats of different ages and sexes. Because the project had started later than planned, he had succeeded in overseeing only one phase of doubling: from twenty rats to forty. Although he didn't know it at the time, this would be the last of his rodent experiments.

He devoted his energy into completing the third project, the assemblage of textual excerpts that would form the basis of the "Global Alerting System." For the purposes of producing a tangible outcome, the goal was a book, whose working title was *Perspectives on Adaptation, Environment, and Population*. He had now collected brief statements of "condensed thought" from one hundred and sixty-two scientists. The contributors were asked to write no more than eighteen hundred words; Calhoun would then pore over the essays, edit them further if necessary, and begin to highlight key terms for indexing. He considered this to be the most important project he had ever embarked upon. However radical and naively ambitious it seemed, Calhoun's Alerting System was not without precedent.

In the late 1930s, H. G. Wells—attempting to transmute success in science fiction to political traction—had envisaged something very similar: a repository of information providing any user with comprehensive and up-to-date information on any given subject matter. A total encyclopedia, constantly updated, organized in such a fashion as to allow rapid access to everything humanity had learned. Wells called it the *World Brain*: "It would not be a miscellany, but a concentration, a clarification, a synthesis." With the nations of Europe poised on the brink of war, Wells presented his World Brain as a means of averting conflict through the exchange of knowledge. He insisted there was nothing utopian about it, that his was "a perfectly sane, sound and practicable proposal." The World Brain would provide a conceptual web that bound all humanity together, a quasi-biological system that "spread like a nervous network, a system of mental control about the globe." He described his project as belonging to "a special sub-section of human ecology, which is a branch of general ecology, which again is a stem in the great and growing cluster of biological sciences." For Wells, human cognition was an extension of human biology, and knowledge was a natural system.[9]

In the mid-1940s, Vannevar Bush, an engineer and science administrator who had overseen the initial phase of the Manhattan Project, published a similar vision for a linked-up encyclopedic storage facility. As the volume of global knowledge had long since exceeded the capacity of a single mind, potentially valuable connections between disparate fields of enquiry were being lost in the widening gaps between ever-more isolated academic specialisms. Some form of external, rapidly accessible repository was called for. Bush proposed the "Memex," a memory index. Wells's World Brain had been a multivolume paper library; Bush's version would be a desk-sized mechanical device containing a highly compressed form of information storage—he envisaged an advanced version of microfilm—where keywords were linked between multiple documents. The information was to be arranged thematically, enmeshed in a multidimensional conceptual order rather than a rigid and linear alphabetical system. In at least its functional operation, Bush's Memex was later recognized as a precursor to the World Wide Web.

In the 1960s, Buckminster Fuller—part visionary designer, part showman—had wowed the world with his geodesic domes, a new form of spherical polyhedron that Fuller had not himself created but for which he cannily secured US Patents. He deployed his "Buckyballs" as both brand identity and metaphor, the real product being a joined-up system of global thinking complete with space-age neologisms like "dymaxion" and "synergetics." Among his myriad schemes, Fuller designed a system for brokering global peace through a "World Game," which was a "game" in the sense that game theorists used the word: a decision-making apparatus for choosing between multiple scenarios. World Game proposed devolving international politics into the hands of the citizenry, an entirely cosmopolitan democracy in which every person could participate, and where decision making was decentralized and arrayed instead around the needs not of individual countries but the entire population of what he called "Spaceship Earth."

As cybernetics and systems theory sought to integrate and organize knowledge into an ecological model, recent advances in microchip technology meant that computing power was finally able to deliver on at least some of their heady claims. At the time, the division between speculative science fiction and science fact was not at all clear. But by the end of the 1970s, such totalizing systems were losing their appeal. As more and more of the promises of the cyberneticians and system theorists failed to deliver usable results—with Bush's Memex granted a retrospective exemption—the excitement around such grand ideas had largely fizzled out, and the comparisons with science fiction, which had once made the work seem so enticing, now made it seem like a stunt.

And so, when Calhoun first conceived of his "prosthetic brain" in the late sixties, the idea of building a complex structure to organize and index vast quantities of information into an interlocking communication network was less radical and entirely more orthodox than it now seems. But things had moved on since the project was conceived, and by the time Calhoun's book appeared, the high point of early 1970s cybernetics and systems theory has passed.

WHEN PUBLISHED in late 1983, the book had a new title: *Environment and Population: Problems of Adaptation*, declaring itself "An Experimental Book Integrating Statements by 162 Contributors." Conventionally dressed in dark green cloth, it didn't look very radical. He introduces the concept on the opening lines: "This is a book. It is not a book. It is a bound set of pages that can be read in sequence starting with the first chapter."[10] But Calhoun invites the reader instead to follow a more complex, choose-you-own-adventure style route through the pages, guided by whichever key terms they find most intriguing. It is a text that can be read in any order; the links between sections are thematic, and, per the guiding principle of the icosahedron-shaped architecture, each linked to twelve others, informed by what Calhoun calls "moods." The book is not linear but three-dimensional, what—at least among information scientists—was already being called a "hypertext."

Texts that subverted linear reading had been attempted before, but they were considered avant garde experimental fiction: Marc Saporta's *Composition No. 1*, first published in French in 1962, and British novelist B.S. Johnson's 1969 work *The Unfortunates* had both been released as a sheaf of unbound sheets, the "novel-in-a-box" that could be read in any order. Their intention had been to emphasize how contingency and juxtaposition altered the meaning of a scene. Calhoun's intention was more structured; he wants to simulate "how the human brain interrelates concepts in a network fashion." To help readers get started, he has highlighted forty-two of the paragraphs as being especially important, believing that reading these will produce a "fusion between the minds of reader and author." One reviewer thought this sufficiently absurd that he took two runs at mocking it: "Reading them does tend to blow the mind . . . fusion of a sort. And, reading 42 isolated paragraphs does produce *con*fusion."[11]

What Calhoun didn't say: this was never meant to be a book. The book was just the model, a demonstration of a principle. The principle was that our species' ongoing survival in a crowded world required a better mechanism for efficient communication. What Calhoun called an alerting system was a means of transferring information urgently in response to a crisis. When he thought back on how it all began, it had been the adrenaline. Adrenalin was the alarm bell. Overpopulation, in rats and in humans, was a threat because it eventually created more social contact than the body could deal with. Unwanted social contact led to the fight-or-flight response being triggered. Repeat that operation enough, it led to stress. Stress built up, depleted the organism's ability to cope. Prevent emigration, exhaust the opportunities for flight, and what emerged was the destructive vortex of the behavioral sink, or slow death by social isolation. The adrenal system was the body's alerting system; it kept the species alive. Calhoun hadn't been trying to write a book, he had been trying to design an adrenal system for humanity.

Critical reception of *Environment and Population* was cautiously favorable. What worked as experimental fiction didn't translate into experimental science. A lengthy review in the journal *Human Ecology* expressed reservations about the result, but admired the effort, and his ambition: "It is easy to criticize someone who tries something new, anyone that attempts to be different! I applaud Calhoun's attempt," the reviewer concluded. "I hope Calhoun will try again, but with a few changes." Jack would not get a second chance.

SINCE HIS teenage years scaling chimneys in search of swifts, he has remained physically fit and limber. His work has often required crouching or crawling over pens or through narrow gates to access the rat enclosures. He has always worked long hours, but with reduced staffing, he frequently now spends twelve or sixteen hours in the lab at a stretch. His relationship with the administration at NIMH is increasingly strained. The ambient bureaucratic scrutiny taxes him, first emotionally, now physically. He begins to suffer periodic bouts of chest pain.

Still, he is active and remains in generally good health. Every year he climbs the deodar cedar outside their house on Cheshire Drive to hang Christmas lights. Among the box of decorations: a small icosahedron he had once made, now a festive bauble. In December 1981, sixty-four years old, he is attaching the lights to the cedar when he loses his footing. Jack the tumbler, tumbling twenty feet against branches to land on the frozen ground of his yard. His injuries amount to nothing more than heavy bruising, but his body is slower to recover.

By late summer, pain in his chest becomes more frequent, spreads to his leg. He is prescribed nitroglycerin for angina, but the medication proves ineffectual and leaves him with severe headaches. Ten days before Christmas of 1982, his car breaks down driving to Poolesville. Outside, it is several degrees below freezing. He is now alone in open country, with no coat. He sets off along the road on foot. When chest pain and exhaustion force him to stop, cold forces him to carry on. It is seven miles before he finds a house from which he can call for help.

Although weakened by the period of exposure, he is adamant the Christmas lights must go up. Less than a week later, and against Edith's wishes, he takes down the box of lights and heads out to the yard. As he reaches the upper branches of the cedar, his breathing becomes labored, a great tightness gathers behind his sternum, and there, perched atop the cedar, he experiences his first heart attack. His note of the episode is terse, understated: "Had trouble getting down."

By February, he is almost unable to walk. Standing a little over five foot three, he loses fourteen pounds. A bypass operation is scheduled for March. The operation is a success, and he is soon back at home, but convalescence is slow. His book complete and no more rodent projects to superintend, he spends much of 1983 recuperating from his surgery. By December he is back at work full-time, but with his experimental work forcibly curtailed, he becomes ever-more disenchanted and pessimistic.

CHAPTER 21

ECCO LIBRIUM

ON a clear October night in 1957, Randall Lowell and his colleagues watched a new star rising over the Pocono mountains in Eastern Pennsylvania. They had stepped outside from a long day at a conference of Hoffmann-La Roche managers feeling triumphant: the drug their team had been working on had just been approved for human trials. Chlordiazepoxide was the first of an entirely novel class, the benzodiazepines. Its discovery had been a fluke. Initially synthesized by Polish American chemist Leo Sternbach in the 1930s, chlordiazepoxide was part of a project to develop artificial dyes. Chemically interesting but useless for its intended purpose, Sternbach stabilized the compound and shelved a vial for posterity, where it gathered dust for two decades.

In the mid-1950s, Hoffmann-La Roche zoology technician Beryl Kappell was clearing out the same shelves and found the vial. Kappell had been testing various compounds for muscle relaxation in animal trials and took a chance on the chlordiazepoxide. To her amazement, it worked. And then some. On her test subjects, chlordiazepoxide also had additional calming effects: reducing aggression and seeming to promote tranquility in anxious and stressed lab animals.[1] If the human trials were successful, chlordiazepoxide promised a chemical cure for anxiety and stress in people. High above the Poconos, the new star tracked past the old constellations smoothly, north-northeast across a clear sky. To a jubilant Randall Lowell, Sputnik seemed a portent heralding a new age: the age of the mood drugs.

Hoffmann-La Roche would bring chlordiazepoxide to market under the trade name Librium in 1960. The name suggested a capacity to correct imbalance, to restore equilibrium. The discovery of chlordiazepoxide's anxiolytic—that is, anxiety reducing—effects triggered a drive to develop new benzodiazepines. These quickly proved remarkably effective for the therapeutic treatment of serious alcoholic withdrawal symptoms, preventing life-threatening seizures and eliminating panic. Three years later, they released another benzodiazepine, *diazepam*, branded as Valium.

America was growing accustomed to chemical cures. During the 1950s, tranquilizers such as *meprobamate*—available as Miltown or Equanil—had been widely prescribed. But Miltown was really more of a sedative than a true anxiolytic, a tranquilizer. Valium was also a gentle sedative, but it was a powerful anti-anxiety medication, more than twice as potent as Librium. By 1969, it had become the country's most prescribed drug, and would retain that position until 1982. In 1978, enough Valium pills were handed out to have allocated ten apiece to every man, woman, and child in the United States: two-point-three billion doses.

Alongside the anti-anxiety medications came the antipsychotics. The first of these, *chlorpromazine*, later marketed as Thorazine, had been discovered in France in 1950, and by the mid-1950s was becoming widespread in clinical practice worldwide. Administered to schizophrenics, Thorazine drastically reduced the expression of violent behavior without the accompanying narcosis. The patients appeared calm and unstressed.

Calhoun's old colleague, Joe Brady, had been testing these very sedatives on his monkeys and rats during their time at Walter Reed. Brady recalled how, in the early 1950s, researchers at a pharmaceutical lab in New Jersey had been trialing the effects of a drug called *reserpine* for cardiovascular research when they noticed an unexpected side effect: "They discovered that when they dosed the rhesus monkeys, generally pretty hostile and aggressive animals, they could put their fingers safely in the monkey's mouth. The monkey was awake but had lost all his aggression."[2] Calhoun recalled how the crates in which the drugged monkeys were held made him squeamish.

Not all drug therapies were the same: Humphry Osmond's use of hallucinogens at Weyburn had been part of an effort to empathetically understand the experience of the mentally ill. The new drugs sought to remove the experience altogether. Even as Calhoun's allies in psychiatry had been trying to restructure the physical environment to alleviate stress and prevent the eruption of violent behaviors, the rise of psychopharmacology was beginning to make their efforts appear redundant. Robert Sommer's carefully arranged seating plans and placing of screens to create a feeling of privacy and safety based on personal space, Kiyoshi Izumi's designs for the Yorkton asylum, Aristide Esser using a territorial model of space to reduce fear and aggression in patients—all these were surely very useful interventions, but Thorazine and reserpine had much the same effect with significantly less effort. Promotional literature trumpeted how "THORAZINE® quickly puts an end to violent outburst." The only design changes now required were fewer restraints and holding cells. Eventually, you wouldn't need institutions at all. "THORAZINE® helps to keep more patients out of mental hospitals," one advert promised. Where would they go? "More patients can be treated in the community."[3] The treatment came to be known as a "chemical straitjacket."[4]

As the American prison population began its vertiginous rise during the 1970s, the availability of institutional psychiatric care continued its steady decline. Both were, in a sense, the result of drugs: incarceration in the penal system rose as a consequence of Nixon's War on Drugs, while the availability of the chemical straitjacket saw inpatient psychiatric placements fall. The numbers of medicated outpatients increased, as did the range of conditions for which the drugs proved effective: Thorazine was soon being indicated for relief of symptoms of menopause in women, and for controlling nervous vomiting in children.[5]

He had seen it coming. Back in 1957, Calhoun had warned his Space Cadets of the increasing pressure they were under to use tranquilizers: "Forget about everything else and give them a goose of this stuff, and they will be happy."[6] He had promised them that their work was more meaningful, while admitting it was less economical. It looked as if the drugs cut the behavior off at the source. But really, what they cut off was its expression. As Calhoun saw it, the drugs did nothing to address the underlying problems that led to social stresses and psychotic violence. They were a quick fix, made a monkey in a metal crate so docile you could put your finger in his mouth, or send him up into space.

AS THE pharmacy of chemical cures expanded, so the importance given to environmental design began to recede. The types of socially engaged, community-based programs Len Duhl had wanted to introduce when he left NIMH for HUD were difficult and they were expensive, requiring vigilant fieldwork and feedback from on-the-ground researchers, extensive training and resources, and a large, skilled, and dedicated network of experts, educators, and administrators. But it wasn't only the expense. While the environmental causes of stress and unhappiness and aggression varied from city to city, even—as the casework of Herb Gans in Boston's West End had shown—from block to block, their biochemical causes remained the same. Biochemistry held a causal priority over the local conditions. Simply put: if you knew the biochemistry, you didn't need to know the environment. Thorazine had the same effect in New York City as it did in Elkton, Tennessee.

The optimism of the late sixties that had made people like Duhl, Osmond, Hall, and Sommer believe that altering physical space could change the way people behaved was being replaced by a weary acceptance of the status quo. Many of the environmental design projects had been abandoned, deprived of funding, or failed to deliver on their early promise.

Meanwhile, where the association with ecology, ethology, and—by extension—scientific rationalism had once been a point in favor of urban revisionists such as Gans and Oscar Newman, the use of biological models was now growing troublesome. During the 1970s, the ecological turn that had been driven by popular accounts such as Ardrey's and Morris's began to roll around again. Attempts to explain human behavior with reference to heredity were increasingly criticized as faulty, reductionist, and racist. In 1975, renowned ant specialist Edward O. Wilson had published *Sociobiology: The New Synthesis*. The book patiently worked through the evolutionary benefits of social structures in a series of animal cases, moving from social behavior in insects—Wilson's specialism—on to fish, birds, and mammals. In its concluding thirty pages, Wilson turned that attention to humans. This final section would prove hugely controversial, as critics saw in Wilson's equivalence between ant societies and human societies a repudiation of human freedom and a justification of social inequality.

What was now being derided as "genetic determinism" came to be linked with noxious ideologies: a 1969 essay by Arthur Jensen on race and IQ reawakened the dormant specter of eugenics, drew lines in the sand. In the ongoing nature-versus-nurture debate, the liberal consensus shifted hard toward nurture, abetted by popular evolutionary biologists such as Richard Lewontin and Stephen Jay Gould, who challenged the "gene's-eye view" of biology and publicly dismantled the scientific credibility of both racial categories and intelligence testing, respectively.

It wasn't that the science became "politicized" so much as the political implications of speaking of human and nonhuman animals were brought to the surface. Calhoun's equivalence of rats and humans had consequences; an account of evolutionary success that held that only the strongest and most dominant individuals would profit was easily mistaken for an endorsement of the idea that might is right. That sort of talk was fine applied to rats, but when scaled up to human societies, the proper disciplinary domain shifted from ethology to politics, and in so doing opened the door to a new form of criticism. There was a sense that those who sought to employ a biological perspective to explain human behavior were tacitly aiming to enforce a particular ideology, embedding an essentialism about human nature that regarded any deviation from biological norms as a deviation from moral norms. To employ ethological models was to corroborate socially conservative views about marriage, motherhood, and hierarchies. What such thinkers really wanted—or so their critics alleged—was to naturalize an account of social arrangements that valorized the nuclear family and collected violent criminals, single mothers, and homosexuals together as "deviants."

After his success in the 1960s, Ardrey, who was close with both Calhoun and MacLean, now found himself increasing embattled. He wrote to Poolesville to warn that he thought "the whole movement of evolutionary thought" was in trouble. "What started as a left-wing attack from certain areas of the social sciences spread rapidly," Ardrey complained, "till by the time it hit Paris my name was coupled with 'fascist.'"[7] Although Calhoun stayed out of the debate around sociobiology, he was aware the tide was turning, noting in his annual report of 1979: "It is fairly dangerous now for a scientist's professional status to espouse that heredity plays a role in social life."[8]

By the late 1970s, comparing people—the urban poor especially—to animals was falling out of favor. The analogies between crowded rats and city slums played into the hands of those who saw the slum dwellers as congenitally flawed and irremediable. Sociologists increasingly resisted "extrapolations from rat pens to cities,"[9] complaining that the sort of animal models proposed by Calhoun and deployed by Ardrey had resulted in a situation where a "red-eyed, sharpfanged obsession about urban life stalks contemporary thought."[10] While Calhoun had great respect for the complexity of rat societies and saw their behavioral derangements as clear evidence of the destructive effects of high population densities, for those who held rats in less esteem it was possible to draw a different lesson from the analogy: the urban poor behaved like rats because they *were* like rats.

"John B. Calhoun cannot be held responsible for metaphors drawn from his classic studies of rat populations," sociologist Eugene Walter allowed, "but they do have a destiny in urban mythology." Warning of the trend toward what he called "vulgar Calhounianism,"[11] Walter added: "I have heard responsible administrators speculating about density ratios as if they expected the tenants to abandon their nests and eat their young." Rather than make a case for the plight of the crowded poor, to invoke Calhoun's rats was to be complicit in their dehumanization.

The type of approach that sought to explain and address social problems as problems of environmental design failing to meet biological needs was coming under increasing attack. Having been overhyped as a magic bullet, it was now criticized as being a distraction from more fundamental issues of underfunding, poverty, and inequality. Meanwhile, diagnosing crowding as a source of social problems was only a short step from blaming the poor for their problems. Criticisms of crowding were easily conflated with accusations of profligate fertility; the reason why the slums were crowded was because their residents had too many children.

The sinister ambiguity of "birth control" meant that opinion had turned on the problem of overpopulation, too. Excessive numbers of humans were once seen as integral to environmental destruction, and the early environmentalists such as Vogt and Osborn had campaigned for contraception as an obvious corollary of conservationism. Unfortunately, but perhaps inevitably, that had led to an emphasis on reducing population growth in developing nations and among the urban—and often non-white—poor, which added the taint of racial discrimination to the already unavoidable antihumanism. If universal misanthropy was bad, selective misanthropy was that much worse.

In response, environmentalism peeled itself away from population control, as what was now styled as "the green movement" sought to sever historical ties to the birth control lobby. Campaign groups such as Greenpeace and Friends of the Earth would make public statements to distance themselves from their former allies in the population control camp and to stress that they didn't blame the world's poor for the world's problems.

One constituency happy to talk about overpopulation were the science fiction writers, who mined the subject for its dystopian possibilities, resulting in a spate of novels gleefully predicting dire consequences from unchecked population growth. Harry Harrison's 1966 *Make Room! Make Room!* was followed by the Hugo Award–winning *Stand on Zanzibar* by John Brunner in 1968. In 1972, Thomas M. Disch's *334* predicted a grim future of compulsory sterilization in a grossly overpopulated New York. *Make Room! Make Room!* would subsequently be used as the basis for the Charlton Heston blockbuster movie *Soylent Green* in 1973, modifying the plot to make surplus humans an apt solution to resource scarcity. Crucially, most of these stories were set in the near future: *Make Room! Make Room!* in 1999, *Stand On Zanzibar* in 2010, *334* in 2025. That association with hyperbolic sci-fi projections made overpopulation seem a purely speculative threat, while simultaneously making it difficult to imagine a response to overpopulation that didn't involve some form of brutally repressive infringement of personal liberties, even—in the case of *Soylent Green*—cannibalism. Worries about a crowded world could stay where they belonged: between the gaudy covers of a sci-fi paperback.

FITTINGLY, THEN, it was a work of fiction that would perhaps do most to communicate the positive message Calhoun had wanted to convey. Journalist Robert Conly had visited Poolesville on assignment with *National Geographic* magazine in 1969. He didn't write an article about Calhoun's laboratory in the end, but nor did he forget the rats. Two years later, under the pen name Robert O'Brien, he would release a children's novel: *Mrs. Frisby and the Rats of NIMH.*

The Rats of NIMH tells the story of a family of mice saved from certain death by a group of wise rats. The rats are escapees from a place called "Nimh," where white-coated scientists performed experiments upon them that resulted in radically advanced brainpower. In the labs are both rats and mice, a mixture of wild-caught stock and albinos. The Rats of Nimh have learned to read and write, they use electrical lighting, understand mechanics, and now use these powers to help Mrs. Frisby relocate her young family.

Conly died shortly after *Mrs. Frisby and the Rats of NIMH* was published in 1973. He would not have seen the culture-inducing Universe 35, nor read of Calhoun's program for creating "super rats." But Calhoun saw so many parallels between the rats in Conly's story and his own work that he was convinced he had been a model. There were details that seemed more than coincidental: the leader of the Rats of Nimh, Nicodemus, was blind in one eye, just like the dominant male in one of Calhoun's early experiments. When ratcatchers from NIMH arrive to collect wild specimens, they speak of "waves" of rats running through the territory, much like the results from Maine forest studies:

> "There won't be another wave. Not tonight. Probably not for four or five nights."
> "Word gets around."
> "You mean they communicate?" A third voice.
> "You bet they communicate."[12]

Conly even made reference to the effects of crowding on behavior, although for the rats, the analogy ran in the other direction: "Sometimes, we were accused of biting human children; I didn't believe that, nor did any of us—

unless it was some kind of a subnormal rat, bred in the worst of city slums. And that, of course, can happen to people, too."

In 1979, Don Bluth, creatively stifled in his position as director of animation at Disney, set up his own production company. Their first feature film, released in 1982, was *The Secret Of NIMH*, an adaptation of Coney's novel. The animated movie brought a new audience to Conly's story, and fresh press attention to Calhoun.

That summer, *Science News* carried a cover story titled "The (Real) Secret of NIMH,"[13] which featured stills from the movie and a profile of Calhoun's work. The story was picked up in *The Washington Post*, too. "Rats! The Real Secret Of NIMH" featured an interview with Calhoun, who was now being described as "veteran NIMH scientist," and hailed his work as prescient: "*NIMH* was another instance in which science fiction, even in a child's story, anticipated science fact."[14] According to the *Post*, Calhoun remembered Conly visiting the lab in the late 1960s, and even suggested that Mrs. Frisby's name was taken from the blue Frisbee hung behind the door of Building 112 ("to help when things got too stressful for us," Calhoun explained).

One of the things Calhoun liked most about *The Rats of NIMH* was that it characterized the intelligent rats as outsiders whose creativity emerged as a response to adapting to challenging circumstance. In his rat pens, he had watched the dominant, high-velocity rats commandeer the majority of the space and restrict access to females. Below them in status, a declining gradient of power. The severely withdrawn were a lost cause: contributing nothing, tending only to themselves. But there was a middle category of introverts, which—although ostracized from much of the social activity—remained curious about their surroundings. Driven out of the breeding colonies, bloodied but unbowed, they "keep probing into all aspects of their environment even though this regularly exposes them to aggressive retribution from dominant associates. They more readily take advantage of changed circumstances."[15] It was among this predominantly homosexual group of rats that he had witnessed the most innovative behaviors. As one respondent had put it, the high-velocity animals had no incentive to become creative because they were "too busy getting their girlfriends, getting the best apartments to live in, getting the best food, to have the drive to develop a creative experience."[16] High-velocity males were the stereotypical high school jocks. By contrast, the introverted but functioning males were driven to explore new strategies for survival.

The dominants were conservative, maintaining the status quo and exhibiting behavior important for survival under environmental conditions that persisted through time. But environments would change, and therefore diversity and innovation were required. At Towson, Calhoun had watched the rats excluded from breeding colonies but unable to emigrate divert their energies into burrow modification instead, rolling balls of mud to form walls and block out drafts: behaviors he had never witnessed before or since. It was these individuals, which he called "creative deviants," that were critical to the future of society:

> Those socially lower status members inhabiting the subordinate half of the habitat provide a safety factor for adapting to changed circumstances. Many of our past studies show that members of such subgroups have a higher probability of developing both deviant and novel behaviors, some of which provide them with a readiness to accommodate to changed circumstances, and thus enhance species survival.

When Calhoun was undertaking his final rat studies, he had encouraged the animals to settle in different sides of the pen, creating a dominant half and a subordinate half. But he stressed that "such differences between groups or subgroups should not be in terms of an equality-inequality gradient"—one was not better than the other: "Rather it is that the two kinds of groups perform different, but equally important, functions for species survival."[17]

With his role within the institutional structure at NIMH increasingly precarious, he cast himself as a lower-velocity individual, marginalized, forced to innovate in order to survive. And in the new mental health agenda, he saw an attempt to suppress such deviants, to make everyone "normal." Believing that a diversity of personality types was essential to maximize the possibilities for creativity and innovation, he worried that the goal of mental health programs was increasingly directed at the homogenization of behavior, treating introversion as a pathology for which medication was the cure. In the rats of Conly's Nimh, he recognized himself: they, too, were creative deviants. By contrast, his superiors were conservative, and their refusal to listen to those lower in the hierarchy—who might help build more effective information networks—did not bode well for the future of society and the species.

A decade after the previous wave of media interest, the reportage this time around did not focus on the apocalyptic consequences of crowding, but on Calhoun's attempts to enlarge the conceptual capacity of his animals. *Science News* claimed that "the most telling resemblance between the actual and fictional rats of NIMH is their culture. Calhoun had written about the idea of creating cultured rats as early as 1967."[18] The *Post* article described how he was conducting "cultural, not chemical" experiments with Norway rats.[19]

But with the falling reputation of environmental design and the availability of the new mood drugs, this late surge of popularity was to prove inadequate to save Calhoun's career.

IN MARCH of 1983, an internal memorandum intended for the ADAMHA-NIMH director William E. Mayer was mistakenly delivered to Calhoun's office. The memo was from Frederick Goodwin, who in 1981 had replaced John Eberhart as head of intramural research and was now Calhoun's immediate superior. Goodwin was writing to Mayer to arrange a meeting to discuss criteria for "Evaluating Scientific Leaders." "As you and I well know," he confided, "one does not attain a leadership role in science without having been thoroughly tested by one's peers." Goodwin spoke of the high levels of drive and energy that were required to maintain that status, of the restless innovation of the scientific leader. While acknowledging the importance of allowing scientists creative autonomy, he noted that "recently, it has been suggested that we ought to be exercising tighter controls over our scientific leaders particularly with respect to the time they spend physically away from their laboratories." He then turned to the issue he wished to meet Director Mayer to discuss:

> There are times, of course, when health fails, energy levels fall with advancing years, or for other reasons the creative flame burns low. When that happens, research productivity suffers. Their standing in the field slips. They are displaced by eager young investigators striving toward the top. At that point steps must be taken to remove the individual from a leadership role. There are several ways to do this. None is easy. But it can be done. It has been done. And it will be done again.[20]

The memo had been composed just two days after Calhoun had announced his imminent heart surgery. Calhoun wasn't named, but the references to failing health, extended absences, and low energy were unambiguous. He was unsure if Goodwin's using the wrong address was a Freudian slip, or a warning. The administrative machinery was beginning to turn, a trap was being set.

Goodwin was a specialist in the study of bipolar disorder, with a particular interest in biological psychiatry. During 1983, he had helped broker a deal with the National Institute of Child Health and Human Development to establish a new Primate Center at Poolesville. Goodwin announced the news that September, explaining that the Primate Center was to be led by "an unusually gifted young investigator"[21] called Stephen Suomi.

Suomi was arriving from Wisconsin, where he had been working with the controversial psychologist Harry Harlow. Harlow's experiments involved removing baby monkeys from their mothers to investigate the consequences of enforced social isolation. Contained within bare metal cages with only a wire-frame "doll" for company, deprived of any physical contact for months at a time, the baby monkeys became aggressive and emotionally withdrawn. They mutilated their own flesh in acts of self-harm. For those who required it, Harlow's experiments provided scientific proof that physical contact, social interaction, and maternal care were important for emotional development. Suomi's work was a follow-up project, using a suite of pharmaceutical therapies to try to rehabilitate the damaged infants. The plan was to create clinically depressed monkeys by enforced isolation, then attempt to repair the damage using the new mood drugs such as chlorpromazine, imipramine, and a recently discovered drug that acted on the serotonin transport system called *fluoxetine*.[22]

As a fellow animal experimenter, he acknowledged the significance of Calhoun's work, but he was dismissive of its application outside rodent studies. As Suomi saw it, while Calhoun had "convincingly demonstrated that the incidence of litter neglect, abandonment, or destruction is directly related to population density in Norway rats," the leap to human equivalents was not warranted: "Current evidence regarding the relation between population density and human child abuse is mixed at best."[23] Suomi maintained that rats were a poor substitute for primates, which were not only closer to humans in physical appearance, but displayed "greater similarity in cognitive capabilities, behavioral repertoires, social groupings, and most importantly for potential models of child abuse, in development of mother-infant and other social relationships." It was easier to observe emotional distress in a monkey than a rat. With his metal crates and single-variable experiments, Suomi would provide a much more predictable model, and—crucially—one that was more suited to drug testing. Suomi's approach fitted the growing interest in biochemistry and neuroscience, and tackled mental health and its treatment as a problem for individuals rather than a consequence of the wider social and physical environment.

Courtesy of the deal Goodwin had struck with the National Institute of Child Health and Human Development, Suomi was bringing a large grant with him, and NIMH would be providing the laboratory space. As of January 1984, the new Primate Center was to occupy Building 112 on the Poolesville campus—Calhoun's URBS lab.

Making the announcement, Goodwin had explained what had been going on behind the scenes: "We already knew of Dr. Jack Calhoun's plans to terminate his experimental programs and after some discussion with Dr. MacLean about his plans, we moved ahead in our negotiations with NICHHD." MacLean's plans were to retire at the end of the year. With Eberhart already gone, Calhoun was losing his remaining protector. Goodwin proudly declared that they were putting in place "the beginnings of a long-term dream of Dr. MacLean's—a legacy, if you will—a viable full-fledged program at Poolesville which will utilize that beautiful facility for its intended purpose of studies of the interrelationships between brain and behavior."[24]

Calhoun was to be evicted. Initially moved to one of the neighboring buildings at Poolesville, he was subsequently relocated to an adjunct facility in Bethesda, where he occupied an office filled with all the data from his last round of experiments, including the twenty-five million data points from the computerized rat environment, as yet unaccessed. Although he had hoped to use the years before his retirement to process the findings, and to write a two-volume book on his life's work, he now had only two remaining staff to assist in what was a mammoth undertaking. The relocations and reorganizations ate into his time.

In December 1983, President Reagan had pledged his administration's "continued support for the goals of the Federal Design Improvement Program."[25] FDIP had been established by Nixon in 1971 and covered all manner of government-funded design—from agency logos to road signs. Reagan announced the establishment of a "Presidential Design Award" to be presented every four years. In October 1984, Calhoun was among the first four recipients, acknowledging his contributions to "Urban Environmental Design."

The following year, he was required to undergo his annual performance assessment by his new boss, Dennis Murphy. Murphy ranked him "minimally satisfactory." In light of his presidential award, and considering the obstacles that had been placed in his way, Calhoun thought this egregious. He replied:

> I thought my performance for the past year merited one of the two top performance ratings (Outstanding or Highly Successful), but that I could put up with "fully successful" if I could just be left alone to complete my writing.[26]

THE CHANGES had been coming for many years. Before Bertram Brown was dismissed as director in 1977, almost two thirds of NIMH's grant funding was awarded to social scientists and psychologists. But with the rise of the mood drugs, that proportion now plummeted. In 1982, Congress even issued a directive that ordered NIMH to cut its support for social research.[27] With the shift away from research on broader psychosocial problems to specific mental "diseases," funding for the Institute bounced back up in the 1980s, but the money was all for biological psychiatry, which promised greater success through a narrower research focus. Rather than social research programs attacking seemingly intractable social problems—unpopular with the more conservative administrations—they would study the underlying biological causes of specific mental disorders and provide new treatments that were less expensive, more direct, and easier to implement.

Calhoun was appalled by this; he complained that NIMH "essentially considers humans as test tubes, vessels in which drugs are introduced and physiological changes observed."[28] Concerned at the direction in which the field of mental health was heading, he reached out to a journalist at *The New York Times*:

> My broader view of what humanity is all about no longer conforms to the 1982 to 1986 policy of the National Institute of Mental Health, i.e., that the only research into mental health justifying financial support is that contributing to the advance of "neuron-science technology."[29]

His assessment wasn't hyperbole. In 1981, shortly after being confirmed as the new director of ADAMHA, William Mayer had presided over an awards ceremony, at which Calhoun had been one of the recipients. Mayer concluded his presentation with a short speech about the direction of research in the field and the bright future for biological psychiatry. At the end, he triumphantly concluded: "Mental Health is drugs, period!"[30]

By 1986, Mayer's statement had largely come true. That May, Calhoun clipped a conference notification: Frederick Goodwin was moderating a debate at the annual meeting of the American Psychiatric Association titled "Resolved: The future clinical uses of neuroscience technology are so profound that research funds should be diverted from psychosocial research."[31] Calhoun annotated the clipping: "No longer is there any reason why we should try to understand how our relations with our fellows derail our ability to make choices to seek fulfillment; 'neuroscience technology' alone knows what people should be, can see that they so become." The following year, fluoxetine—one of the new drugs Suomi had been testing on his emotionally wounded monkeys—would receive final FDA approval. It appeared on the market a few months later under the trade name Prozac.

Driven out of his lab, his research increasingly obsolete, Calhoun had run out of space. There was an ironic symmetry to the turn of events. After all, it had been the failure of a chemical fix—the short-lived miracle of Curt Richter's ANTU—that had first spurred the Rodent Ecology Project to investigate the environmental mechanisms of population control and triggered Calhoun's lifelong fascination with the impact of the environment upon behavior. Now, a new chemical fix was displacing the style of research he had done so much to promote.

Calhoun submitted his resignation to Goodwin on July 30, 1986. Two weeks later, Goodwin had still not replied. "Why should he?" Calhoun bitterly wrote. "1986 is '1984.' C'est finis."

CODA

QUIETUS

HEART trouble runs in the family. Jack's father, James, died of a heart attack. His brother had a double bypass operation, his sister, a quadruple. With just the one bypass surgery, Jack has gotten off lucky. But he knows what is coming.

He spends his retirement helping Edith trace their family histories, making trips to Ohio and Tennessee to trawl archives for genealogical data. The work absorbs him, an outlet for his habits: the branching lineages of their ancestry are another sort of pattern, another way of organizing the world. Tracing a population that does not grow but instead winnows and recedes, shrinks into colonial wagon trails, sails back east, dissipates in the signal void of Scottish clans.

There is bitterness, too. The circumstances of his departure from NIMH grate upon him, a moral injury. Before leaving, he was unable to access the computer reels containing the data from his final rat study: the location and movement of every individual during the four years up to the abrupt termination of Universe 35. But outside the long-since-dismantled computer system at Poolesville, the information is inaccessible, mute black spools of tape. He is frustrated by the unfinished work.

Jack keeps busy, gives talks on his work at the Family Center of Georgetown University Medical School. Mostly, he holes up in his basement office, a burrow under the house, going back over his life's work. He organizes folders thematically, makes clippings. He indexes, chooses new keywords, indexes it again. The different versions tell the story of his life a hundred different ways. In the margins, and between the clippings and photocopies he pastes into his notebooks, his flowing cursive begins to snag, the ligatures break apart. By the 1990s, he is printing his letters in an increasingly labored hand.

In early summer of 1995, Edith notices that Jack has been collecting favorite books and articles from various shelves around the house, bringing them down to his basement office. Taking papers to his burrow. Late August, Edith and Jack plan a road trip through New England to Maine, travelling back through Vermont and New Hampshire. Retracing the route from Bar Harbor. They are thinking of selling up, moving north to a smaller home for their retirement years. Edith stands by the car as Jack locks up the house. He pauses.

"I won't be back," he says.

On their return journey they cross the Connecticut River from Vermont into New Hampshire, stop at an inn for the night. In the early hours of Labor Day morning, Jack has his second heart attack. He's taken to the hospital in Hanover. Although conscious and able to speak in the ambulance, he later slips into a coma from which he will not awake. Life support is switched off on Thursday, September 7.

EXACTLY SIXTY years earlier, just eighteen years old, Jack had published a short piece in *The Migrant*, the journal of the Tennessee Ornithological Society.[1] It's a descriptive report from a weekend spent birdwatching at a thirty-acre highland marsh, some seven miles north of Springfield. Despite his youth, he seems to know every tree and plant, every bird by call and sight: the wood duck and crested flycatcher, the chickadee, titmouse, and the elusive mourning warbler. During his stay, a red-shouldered hawk is a constant, tracing wide loops high above. He sleeps in an abandoned cabin, the place suiting him fine: "That night I was entertained by the peepings of the tree frogs, the resonant bellowings of the bull frogs and the distant laughing call of the Barred Owl."

In the same edition, the editors of *The Migrant* regretfully announce that "one of our youngest members," John B. Calhoun, will be leaving for university that fall. There's even a photograph: white shirt open, one hand on his hip, a halfway smile. Muscle Jack, recently voted Most Likely to Succeed. His first published photograph, it's one of the very last of him clean-shaven; once at college, he wore a moustache or goatee beard for the rest of his life. "'Jack' is endowed with tireless energy and enthusiasm and lets no obstacles stop him," the editors' note says. "We will miss him."

As the light begins to fail, Jack knows he cannot stay. Even as he is leaving, he lists the nests he can see, counts the woodpecker holes in the dead trees. He ends his report:

> Reluctant though I was to leave this fertile field of observation, a glance at my watch told me I must be on my way. As I left, a Red-shouldered Hawk circled overhead to bid me a final farewell.

ENDNOTES

CHAPTER 1: A NEW WORLD

1 Andrew White, *A Briefe Relation of the Voyage Unto Maryland* (1634), Maryland Historical Society Fund Publication, 35 (1984): 9.
2 Edward C. Papenfuse, "Thomas Poppelton: The Map that Made Baltimore," online at rememberingbaltimore.
3 Daniel Coit Gilman, "Inaugural address, February 22 1876." Box: 1–5, Folder: 18. Johns Hopkins University collection of university-related ceremonies, speeches, and public events, COLL-0004. Special Collections.
4 Online at planning.baltimorecity.gov
5 Edmund Ramsden, "Rats, Stress and the Built Environment," *History of the Human Sciences* 25.5 (2012): 123–147; 127.
6 Curt P. Richter, "Experiences of a Reluctant Rat Catcher: The Common Norway Rat—Friend or Enemy?" *Proceedings of the American Philosophical Society* 112.6 (Dec. 9, 1968): 403–415; 409.
7 Thomas Jefferson, "letter to Benjamin Rush, 23 September 1800," in *The Papers of Thomas Jefferson, vol. 32, 1 June 1800–16 February 1801*, ed. Oberg, Barbara B. (Princeton: Princeton University Press, 2005), 166.
8 Report from September 20, 1892, qtd. in Garrett Power, "Apartheid Baltimore Style: The Residential Segregation Ordinances of 1910–1913," *Maryland Law Review* 42 (1982): 289–328, 290.
9 Janet E. Kemp, *Housing Conditions in Baltimore: Report of a Special Committee* (Baltimore Association for the Improvement of the Condition of the Poor and the Charity Organization Society, 1907), 21.
10 Philippeaux, M. "Note sur l'extirpation des capsules surrénales chez les rats albinos," *Comptes Rendus Hebdomadaires des Seances de l'Academie des Sciences* 43 (1856): 904–906.
11 Garrett Power, "Apartheid Baltimore Style: The Residential Segregation Ordinances of 1910–1913," *Maryland Law Review* 42 (1982): 289–328, 298.
12 Milton J. Greenman and F. Louise Duhring, *Breeding and Care of the Albino Rat for Research Purposes* (Philadelphia: Wistar Institute, 1923), 11.
13 Helen Dean King, *Studies on Inbreeding* (Philadelphia: Wistar Institute, 1919), 3, 4, 4, 6.
14 Charles Darwin, *The Variation of Animals and Plants Under Domestication*, vol. 2 (London: John Murray 1868), 115–116.
15 Bonnie Tocher Clause, "The Wistar Rat as a Right Choice: Establishing Mammalian Standards and the Ideal of a Standard Mammal," *Journal of the History of Biology* 26.2 (1993): 329–349.
16 Henry Hubert Donaldson, *The Rat: Reference Tables and Data* (Philadelphia: Wistar Institute, 1915), 6.
17 James T. Todd and Edward K. Morris, "The Early Research of John B. Watson: Before the Behavioural Revolution," *The Behavior Analyst* 9.1 (1986): 71–88, 73.
18 City of Baltimore, *One Hundred and Twenty-ninth Annual Report of the Department of Health* (Baltimore City Health Department, 1943), 321.

CHAPTER 2: JOHNS HOPKINS

1 Curt P. Richter, "Experiences of a Reluctant Rat Catcher: The Common Norway Rat—Friend or Enemy?" *Proceedings of the American Philosophical Society* 112.6 (Dec. 9, 1968): 403–415, 403.
2 John B. Watson, "Psychology as the Behaviorist Views It," *Psychological Review* 20 (1913): 158–177, 158.
3 Adolf Meyer Archive Series; Collection MeyA I/3974/9; Alan Mason Chesney Medical Archives, Johns Hopkins.
4 John B. Watson and Rosalie Rayner, "Conditioned Emotional Reactions," *Journal of Experimental Psychology* 3.1 (1920): 1–14.
5 Meyer, letter to Goodnow, Sept. 29, 1920; qtd. in Andrew Scull and Jay Schulkin "Psychobiology, Psychiatry, and Psychoanalysis: The Intersecting Careers of Adolf Meyer, Phyllis Greenacre, and Curt Richter," *Medical History* 53.1 (2009): 5–36.
6 Andrew Scull and Jay Schulkin, "Psychobiology, Psychiatry, and Psychoanalysis: The Intersecting Careers of Adolf Meyer, Phyllis Greenacre, and Curt Richter," *Medical History* 53.1 (2009): 5–36.
7 Mark Suckow, Steven Weisbroth, and Craig Franklin, eds., *The Laboratory Rat* (London: Elsevier/Academic Press, 2005).
8 Curt P. Richter, "It's a Long, Long Way to Tipperary, the Land of My Genes," in *Leaders in the Study of Animal Behavior: Autobiographical Perspectives*, ed. Donald A Dewsbury (Lewisburg: Bucknell UP, 1985), 372, 377.
9 Richter, "Reluctant Rat-catcher," 404.
10 Paul Rozin, "Curt Richter: The Compleat Psychobiologist," in *The Psychobiology of Curt Richter*, ed. Elliott M Blass (Baltimore: York Press, 1976), xxv.
11 Donald Fleming, "Walter B Canon and Homoestasis," *Social Research* 51.3 (1984): 609–640; 611, 614.
12 Kathy L. Ryan, "Walter B. Cannon's World War I Experience: Treatment of Traumatic Shock Then and Now," *Advances in Physiology Education* 42 (2018): 267–276.
13 Walter B. Cannon, *Bodily Changes in Pain, Hunger, Fear and Rage* (New York: Appleton and Company, 1929 [1915]), 217.
14 Walter B. Cannon, *The Wisdom of the Body* (New York: Norton, 1932), 22.

CHAPTER 3: BALTIMORE

1 Curt P. Richter, "Total Self Regulatory Functions In Animals and Human Beings," *The Harvey Lectures 1942–1943* (Lancaster, PA: Science Press Printing Company, 1943), 91.
2 Arthur L. Fox, "The Relationship Between Chemical Constitution and Taste," *Proceedings of the National Academy of Sciences of the United States of America* 18.1 (1932): 115.
3 H. Cardullo, Holt L.J., "Ability of Infants to Taste PTC: Its Application in Cases of Doubtful Paternity," *Proceedings of the Society of Experimental Biology and Medicine* 76 (1951): 589–592.
4 Richter, "Reluctant Rat-catcher," 404, 408, 406, 408.
5 Rockefeller Foundation, "Rats in War and Peace," *Confidential Monthly Report, Trustees* no. 100 (Feb. 1, 1948), Rockefeller Archive Center.
6 City of Baltimore, "Report of the Commissioner of Health," *One Hundred and Twenty-Ninth Annual Report of the Department of Health* (1943), 39.
7 United States Public Health Service, *Keep 'Em Out*, Film (1942).

8 "Anti-Rat Office Here Is Selected," *Baltimore Sun* (June 5, 1943), 24.
9 Christine Keiner, "Wartime Rat Control, Rodent Ecology, and the Rise and Fall of Chemical Rodenticides," *Endeavour* 29.3 (2005): 119–125.
10 Justus C. Ward, "Rodent Control with 1080, ANTU, and Other War-Developed Toxic Agents," *American Journal of Public Health* 36 (1946): 1427–1431.
11 Frank S. Lisella, Keith R. Long, and Harold G. Scott, "Toxicology of Rodenticides and Their Relation to Human Health," *Journal of Environmental Health* 33.4 (1971): 361–365.
12 John J. Christian, "In Memoriam: David E. Davis, 1913-1994," *The Auk* 112.2 (1995): 491, 492.
13 John J. Christian and David E. Davis, "The Relationship between Adrenal Weight and Population Status of Urban Norway Rats," *Journal of Mammalogy* 37.4 (1956): 476.

A STEEPLE FULL OF SWALLOWS

1 Katherine A. Goodpasture, "In Memoriam: Amelia Rudolph Laskey," *The Auk* 92 (1975): 252–259.
2 Ben Coffey, "Swift Banding in the South," *The Migrant* 9.4 (1938): 82.
3 John B. Calhoun, "1938 Swift Banding at Nashville and Clarksville," *The Migrant* 9.4 (1938): 77–81.
4 Edward O. Wilson and Charles D. Michener, "Alfred Edwards Emerson, 1896–1976," *Biographical Memoirs* (Washington, DC: National Academy of Sciences, 1982), 164.
5 Ben Coffey, "Winter Home of Chimney Swifts Discovered in Northeastern Peru," *The Migrant* 15.3 (1944): 37–38; Albert F. Ganier, "More About the Chimney Swifts Found in Peru," *The Migrant* 15.4 (1944): 39–41.

CHAPTER 4: RODENT ECOLOGY PROJECT

1 David E. Davis, "The Characteristics of Rat Populations," *The Quarterly Review of Biology* 28.4 (1953): 373–401, 373, 395, 395, 399.
2 David E. Davis, "The Rat Population of New York, 1949," *American Journal of Epidemiology* 52.2 (1950): 147–152.
3 William Henry Burt, "Territoriality and Home Range Concepts As Applied To the Mammals," *Journal of Mammalogy* 24.3 (1943): 346-352.
4 David E. Davis, "The Characteristics of Global Rat Populations," *American Journal of Public Health* 41 (1951):158–163, 161.
5 John B. Calhoun, "What Sort Of Box?" *Man-Environment Systems* 3.1 (1973): 3–30, 6.
6 John J. Christian, "Endocrine Adaptive Mechanisms and the Physiologic Regulation of Population Growth," in *Physiological Mammalogy, vol. 1.*, eds. William H. Mayer and Richard G. Van Gelder (New York: Academic Press, 1963), 189–353, 303.
7 Calhoun, "What Sort Of Box?" 6.
8 David E. Davis, "The Role of Infraspecific Competition in Game Management," *Transactions of the Fourteenth North American Wildlife Conference* (Washington, DC: Wildlife Management Institute, 1949), 231.
9 Stuart O. Landry, "Obituary: John Jermyn Christian: 1917–1997," *Journal of Mammalogy* 79.4 (1997): 1432–1439.
10 Hans Selye, "A Syndrome by Diverse Nocuous Agents," *Nature* (July 4, 1936): 32.
11 Cannon, *Bodily Changes*, 49.
12 Davis, "Characteristics of Rat Populations," 399.

CHAPTER 5: TOWSON

1 Calhoun, autobiography, National Library of Medicine [NLM] Archive, John B. Calhoun Papers, National Institute of Health..
2 John B. Calhoun, *The Ecology and Sociology of the Norway Rat* (Bethesda, MD: Public Health Service, 1963), 949, 50, 211, 213, 213, 211, 216, 155, 216, 256, 256, 284, 203.

CHAPTER 6: MAXIMUM HUMAN PROTOPLASM

1 Eugene Rochow, "Chemistry Tomorrow," *Chemical and Engineering News* 27.21 (1949): 1510–15; 1511, 1514, 1512.
2 Waldemar Kaempffert, "Chemistry Offers a Way Out for a 'Plundered Planet' in the World of Tomorrow," *The New York Times* (May 15, 1949) E9; editorial, "Mechanical Dreamland," The *New York Times* (May 15, 1949), E8.
3 Fairfield Osborn, *Our Plundered Planet* (Boston: Little, Brown, 1948), 40.
4 Calhoun, *Ecology and Sociology of the Norway Rat*, 185.
5 Qtd. in John B. Calhoun, "the Social Aspects of Population Dynamics," *Journal of Mammalogy* 33.2 (1952): 139–159, 151.
6 Elizabeth Gordon, "How to Have a Private Estate on 105 'by 103,'" *House Beautiful* 90.12 (Dec. 1948): 154–161; 155, 156, 156.
7 Calhoun, letter to J. P. Scott, Dec. 15, 1948. [NLM Archive].
8 "Model of Housing Displayed at Fair," *The New York Times* (May 5, 1939): 47.
9 H. B. Wilson, "Urban Redevelopment as Exemplified by 'Stuyvesant Town' in New York City," US Public Roads Administration, 1945, 2, 4.
10 "Housing Plan Opposed: 'Walled City for Privileged' Is Seen by Union Council," *The New York Times*, (May 27, 1943); Henry S. Churchill, "Met Gits The Mostest," letter to editors, *Architectural Forum* (June 1943).
11 Lewis Mumford, "Prefabricated Blight," *The New Yorker,* Oct. 30, 1948, 70, 72; "The Sky Line: Stuyvesant Town Revisited," *The New Yorker*, Nov. 19, 1948, 65.
12 Robert Moses, letter to editors, and Lewis Mumford, "Stuyvesant Town Revisited," *The New Yorker* (Nov. 27, 1948): 60, 63
13 "New York: New Nightmares for Old?" *Time* (Dec. 13, 1948).
14 Calhoun, letter to Paul Scott, Dec. 15, 1948.
15 John Q. Stewart, "Concerning 'Social Physics,'" *Scientific American* 178.5 (1948): 20–23, 20.
16 Calhoun, autobiography [NLM Archive].
17 Calhoun, "The Social Aspects of Population Dynamics," *Journal of Mammalogy* 33.2 (1952): 139–159, 143, 143.
18 Calhoun, "An Overliving Population That Didn't," 1973, unpublished [NLM Archive].
19 Calhoun, "What Sort Of Box?" 5, 12.

CHAPTER 7: BAR HARBOR, WALTER REED

1 John Paul Scott, and John L. Fuller, "A School For Dogs," *Dog Behavior: The Genetic Basis* (Chicago: University of Chicago Press, 1963).
2 Calhoun, "Remarks Concerning the Orientation of Approaches Within the General Field of Behaviour," prepared for the conference with Hamilton Station Staff, Oct. 1949.
3 In 1947, he had given a seminar at Johns Hopkins School of Hygiene and Public Health titled "Rat-City Studies" and arranged for the delegates to visit his pen. Box 21 part 1, 11.

4 "Rare Research Team Studies Rat Homelife," *The Baltimore Evening Sun* (May 25, 1948), 24.
5 "Scientific 'Rat City' Being Destroyed; Its King Had Harem and 56 Offspring," *The Baltimore Evening Sun* (May 2, 1949), 33.
6 Calhoun, letter to Richter, Aug. 10, 1949.
7 Calhoun letter to Richter, Jan. 17, 1949.
8 They were: John A. Clausen, Carl L. Larson, and W.T.S. Thorp.
9 Calhoun, "The Social Use of Space" (1963) in *Physiological Mammalogy*, eds. Mayer, W. and Van Gelder, R. (Academic Press, New York, 1964), 81.
10 "Millions and Millions of Mice," *Time*, Aug. 3, 1942.
11 Charles Elton, *Voles, Mice, and Lemmings: Problems in Population Dynamics* (Oxford: Clarendon Press, 1942), 215.
12 Charles Elton and Mary Nicholson, "The Ten-year Cycle in Numbers of the Lynx in Canada," *Journal of Animal Ecology* 11.2 (1942): 215–244.
13 Calhoun, North American Census of Small Mammals, Compilations of Research Data, Rodent Ecology Project, Johns Hopkins (1948–1957).
14 John M. Caldwell, Stephen W. Ranson, and Jerome G. Sacks, "Group Panic and Other Mass Disruptive Reactions," *United States Armed Forced Medical Journal* 2.4 (1951): 541–567; 556, 543, 560.
15 Stephen W. Ranson, "The Normal Battle Reaction: Its Relation to Pathologic Battle Reaction," *Bulletin of the U.S. Army Medical Department*, supplemental (1949): 3–11.
16 US Government, *Army Medical Department Research and Graduate School* Pamphlet (Washington, DC: Army Medical Center, 1949), 10.
17 Roy R. Grinker and John P. Spiegel, *War Neuroses in North Africa: The Tunisian Campaign, January–May 1943* (New York: Macy Foundation, 1943), 136; qtd. in Theodore M. Brown, "'Stress' in US Wartime Psychiatry: World War II and the Immediate Aftermath," in *Stress, Shock, and Adaptation in the Twentieth Century*, Cantor and Ramsden (NY: University of Rochester Press, 2014).
18 Kurt Goldstein, "On So-Called War Neuroses," *Psychosomatic Medicine* 5.4 (1943): 376–383, 376.
19 William C. Menninger, "Psychiatric Experience in the War, 1941–1946," *American Journal of Psychiatry* 103 (1946–47): 579–80, qtd. in Brown, 2014.
20 Roy R. Grinker and John P. Spiegel, "The Management of Neuropsychiatric Casualties in the Zone of Combat," in *Manual of Military Neuropsychiatry*, eds. Solomon and Yakovlev (Philadelphia: Saunders 1944), 517.
21 Surgeon General, Department of the Army, Proceedings of the Fourth Annual Conference of Psychologists in the Army Medical Service, 1961.
22 Joseph V. Brady, "Journal Interview 74: Conversation with Joseph V. Brady," *Society for the Study of Addiction* 100 (2005): 1805–1812, 1806.
23 Calhoun, "Job Description for Position at Walter Reed Army Medical Center," 1951 [NLM Archive].
24 Calhoun, "What Sort Of Box?" 13.
25 Calhoun, "Remarks on the Organization of a Population Behavior Laboratory," April 1949 [NLM Archive].
26 Calhoun, autobiography, 1965 [NLM Archive], 29.
27 Calhoun and William L. Webb, "Induced Emigrations among Small Mammals," *Science* 117.3040 (1953): 358–360.
28 Calhoun, letter to William Webb, July 1, 1953.

29 Calhoun, "The Social Use of Space," 27, 76.
30 Charles Darwin, "Struggle for Existence," *The Origin of Species*, 6th. Ed. 1872 (New York: Modern Library/Random House, 1998), 104.
31 Calhoun, "Development of the Concept of Optimum Group Size: Excepts from Publications (with a few annotations)," *SOBS* Doc 42, 30 June 1982—annotation to "Social Use of Space" (1963), 31.
32 Calhoun, letter to William Webb, July 2, 1953; William Webb, letter to Calhoun, June 22, 1953.
33 Calhoun, autobiography, 1965 [NLM Archive].
34 Rioch, qtd. in "Recommendation of Promotion of John B. Calhoun from GS-14 to GS-15," NIMH internal memorandum, March 22, 1955.
35 Calhoun, "What Sort Of Box?" 13, 14.
36 Calhoun, letter to JP Scott, Jan. 12, 1954.

CHAPTER 8: CASEY'S BARN

1 Maryland Historical Trust State Historic Sites Inventory Form, Farm 245, M:21–183.
2 "Animal Buildings Nearing Completion," *NIH record* (Feb. 9, 1953): 5.3.
3 Robert C. Cook, "The Population Reference Bureau," *Science* 118.3074 (1953): 3.
4 W. H. Schneider, "The Model American Foundation Officer: Alan Gregg and the Rockefeller Foundation Medical Divisions," *Miverva* 41 (2003): 155–166.
5 Calhoun, letter to Gene Gressley, July 15, 1983.
6 Calhoun, "A 'Behavioral Sink,'" in *Roots of Behavior: Genetics, Instinct, and Socialization in Animal Behavior*, ed. Eugene Bliss (New York: Hafner, 1962): 298, 303, 303.
7 This represents one of the first recorded uses of the term *pansexuality*.
8 Calhoun, "Population Density and Social Pathology," *Scientific American* 206.2 (1962): 139–148, 146, 146.
9 Philip G. Cox, et al., "Functional Evolution of the Feeding System in Rodents," *PLoS ONE*. 7.4 (2012): e36299.
10 Calhoun, "A 'Behavioral Sink,'" 303.
11 Calhoun, "Population Density and Social Pathology," 146.
12 Calhoun, "A 'Behavioral Sink,'" 303.
13 Calhoun, "What Sort Of Box?" 16.
14 "Honored," *The Sentinel Montgomery County* (Nov. 25, 1982): A–5.
15 NIMH Internal Memorandum, David Shakow Recommendation for Promotion of Calhoun, Feb. 27, 1959 [NLM Archives].

CHAPTER 9: OUT OF THE SINK

1 Bruce V. Lewenstein, "The Meaning of 'Public Understanding of Science' in the United States After World War II," *Public Understanding of Science* 1.1 (1992): 45–68.
2 Stephen Turner, "What Is the Problem with Experts?" *Social Studies of Science* 31.1 (2001): 123–149.
3 Calhoun, "Population Density and Social Pathology," 148.
4 Joe Flower, "Building Healthy Cities: Excerpts from a Conversation With Leonard J. Duhl," *Healthcare Forum Journal* 36.3 (1993).
5 Calhoun, "Origin of the Space Cadets" [NLM Archives].
6 Calhoun, "Residency, 1954–1965," Memorandum, May 5, 1965 [NLM Archives].
7 Conference on the Physical Environment as a Determinant of Mental Health. Held

at APA, May 28–9, 1956. [Hereafter such conferences designated "Space Cadet minutes."]

8 Space Cadet minutes, May 6–7, 1957.

9 John J. Christian, letter to Edward T. Hall, Oct. 14, 1963; Edward T. Hall Papers, University of Arizona.

10 Calhoun, "A Method for Self-Control of Population Growth among Mammals Living in the Wild," *Science* 109.2831 (1949): 333–335.

11 David E. Davis, "Early Behavioral Research on Populations," *American Zoology* 27 (1987): 825–837, 827.

12 John J. Christian and H.L. Radcliffe, "Shock Disease in Captive and Wild Mammals," *American Journal of Pathology* 28.4 (1952): 725–737.

13 John J. Christian, "Phenomena Associated with Population Density," *Proceedings of the National Association of Science: Anthropology* 47 (1961): 428–449, 435.

14 John J. Christian, Vagn Flyger, David E. Davis, "Factors in the Mass Mortality of a Herd of Sika Deer, *Cervus Nippon*," *Chesapeake Science* 1.2 (1960): 79–95; 79, 82, 84, 95, 79, 94.

15 Calhoun, *$R_x$evolution: The Prescription For, or Design of, Evolution*, unpublished manuscript (1974), Section 1–5.

CHAPTER 10: PERSONAL SPACE

1 Harry S. Truman, Inaugural address, speech at the Capitol, Washington, DC (Jan. 20, 1949).

2 Edward T. Hall, *An Anthropology of Everyday Life* (New York: Anchor-Doubleday, 1992), 76.

3 Hall, "The Anthropology of Manners," *Scientific American* 192.4 (1955): 84–91, 86.

4 Hall, "Proxemics—The Study of Man's Spatial Relations," in *Man's Image in Medicine and Anthropology: Monograph IV*, ed. Iago Gladston (New York: International Universities Press, 1963), 422–45.

5 Hall, 1955, "The Anthropology of Manners," 85.

6 David Katz, *Animals and Men* [1937], trans. Hannah Steinberg and Arthur Summerfield (Harmondsworth: Penguin, 1953), 99.

7 Robert Sommer, letter to Hall, Feb. 8, 1960. In Edward T. Hall Papers, University of Arizona, box 15.

8 Hall, letter to Sommer, Feb. 18, 1960. Hall Papers, box 15.

9 Sommer, letter to Hall, Feb. 1964. Hall Papers, box 15.

10 William Henry Burt, "Territoriality and Home Range Concepts as Applied to Mammals," *Journal of Mammalogy* 24.3 (1943): 346–352, 351.

11 Hall, letter to Calhoun, June 15, 1960. Hall Papers, box 15.

12 Edward T. Hall, *The Hidden Dimension* (New York: Doubleday, 1966), 165, 129, 165, 165, 186, 129.

13 Tom Wolfe, *The Pump House Gang* (New York: Farrar/Bantam, 1968), 231.

14 Hunter S. Thompson, letter to Tom Wolfe, April 21, 1968, in *Fear and Loathing in America: The Gonzo Letters, Volume II. 1968–1976*, ed. Douglas Brinkley (New York: Simon and Schuster, 2000).

15 Hall, *Hidden Dimension*, 167–168.

16 Hall, "Human Needs and Inhuman Cities," in *The Fitness of Man's Environment* (Washington, DC: Smithsonian Institution Press, 1968), 164.

17 Robert Ardrey, *African Genesis* (London: Harper Collins, 1961), 30.

18 Ardrey, *The Social Contract* (London: Collins, 1970), 183–184, 217–218, 218, 219.
19 "The Decade's Most Notable Books," *Time*, Dec. 26, 1969.
20 Stephen Moss, "We'd be better off if women ran everything," interview with Desmond Morris, *The Guardian* (Dec. 18, 2002).
21 Desmond Morris, *The Naked Ape: A Zoologists Study of the Human Animal* (New York: McGraw-Hill, 1967), 85, 178–179, 187.
22 For extended discussion of Ardrey and Morris's influence, see Erika Milam, *Creatures of Cain: The Hunt for Human Nature in Cold War America* (New Haven: Yale UP, 2019).

CHAPTER 11: ASYLUM

1 Abram Hoffer, "Treatment of alcoholism with psychedelic therapy," in *Psychedelics: The Uses and Implications of Hallucinogenic Drugs*, eds. Aaronson and Osmond (London: Hogarth, 1971).
2 Aldous Huxley, *Brave New World* (London: Vintage, 1932), 46.
3 Humphry Osmond, "Function as the basis of psychiatric ward design," (1957) in *Environmental Psychology: Man and His Physical Setting*, eds. Proshansky, Ittelson, and Rivlin, eds, (New York: Holt, 1970), 561.
4 Osmond, letter to Aldous and Laura, July 14, 1956, in Bisbee, et al., *Psychedelic Prophets: The letters of Aldous Huxley and Humphry Osmond* (Montreal: McGill-Queen's UP, 2018).
5 Osmond, "Function as the basis of psychiatric ward design," 560, 563–565.
6 Izumi, "LSD and Architectural Design," in *Psychedelics*, eds. Aaronson and Humphry Osmond (Hogarth Press: London, 1970), 387, 395.
7 Robert Sommer, "The Ecology of Privacy," (1966), in *Environmental Psychology: Man and His Physical Setting*, eds. Proshansky, Ittelson, and Rivlin (New York: Holt, 1970), 256.
8 Osmond, "Function as the Basis of Psychiatric Ward Design," 561.
9 Sommer, *Personal Space* (Englewood-Cliffs, NJ: Prentice Hall, 1969), 14.
10 Osmond, "A comment on some uses of psychotomimetics in psychiatry," Hall Papers, EDRF Archives, Kansas University.
11 Kiyoshi Izumi, "LSD and Architectural Design," in *Psychedelics: The Uses and Implications of Hallucinogenic Drugs*, eds. Aaronson and Osmond (Hogarth Press: London, 1970), 382, 389.
12 Osmond, letter to Huxley, July 29, 1957, in *Psychedelic Prophets: The letters of Aldous Huxley and Humphry Osmond*, eds. Bisbee et al. (Montreal: McGill-Queen's University Press, 2018), 345.
13 Charles Goshen, "A Review of Psychiatric Architecture and the Principles of Design," in *Psychiatric Architecture*, ed. Charles Goshen (Washington, DC: The American Psychiatric Association, 1959), 1–6 (1).
14 Humphry Osmond and Bernard Aaronson, "Psychedelics and the Future," in *Psychedelics: The Uses and Implications of Hallucinogenic Drugs* (Garden City, NY: Anchor-Doubleday, 1970), 467, 469–470, 463.
15 Aristide H. Esser et al., "Territoriality of Patients on a Research Ward," in *Biological Advances in Psychiatry, vol. 7*, ed. Joseph Wortis (New York: Plenum, 1965), 37–44, 43, 43.
16 Esser, "Social Contact and the Use of Space in Psychiatric Patients," Abstract, AAAS Meeting, 1965, EDRF Papers, Kansas, Box 54, S.1692.
17 Osmond, "Function as the Basis of Psychiatric Ward Design," 561.
18 Izumi, "Some Architectural Considerations in the Design of Facilities for the Care and

Treatment of the Mentally Ill," *American Schizophrenic Association Journal* (Summer 1967).

19 Sommer, *Personal Space: The Behavioral Basis of Design* (Englewood Cliffs, NJ: Prentice-Hall, 1969), 23.

20 Esser, *Preface to Behavior and Environment: The Use of Space by Animals and Men* (New York: Plenum, 1971).

21 Hall, "Environmental Communication," in *Behavior and Environment: The Use of Space by Animals and Men*, ed. Esser (Plenum: New York, 1971), 260.

CHAPTER 12: PENITENTIARY

1 US District Court for the District of Columbia; Leonard Campbell, John McIlwain, Richard Kinard, Eligah Hair Smith, plaintiffs, vs. Charles M. Rodgers, Superintendent, DC Jail, DC, Kenneth Hardy, Director, DC Departments of Collections. DC, Walter R. Washington, Mayor and Commissioner of the DC, Defendants; Complaint for injunction, declaratory judgment and other appropriate relief.

2 Richard Nixon, "Towards Freedom From Fear," *Congressional Record—Senate* 114.10 (1968): 12936–39, 12939.

3 Ronald Goldfarb, "No room in the jail," *The New Republic* (March 5, 1966): 12–14, 12, 12.

4 Ardrey, The Social Contract, 233.

5 Ben H. Bagdikian, "A Human Wasteland in the Name of Justice," *The Washington Post*, Jan. 30, 1972.

6 William L. Claiborne, "Experts take a look at—and smell of—the District Jail," *The Washington Post*, Oct. 17, 1971.

7 Calhoun, "Remarks about a visit to the D. C. Jail," Jan. 18, 1972, URBS doc 194, Part B.

8 Calhoun, "Brief Anthology on 'Confinement,'" Nov. 19, 1971, URBS doc 194, Part C.

9 "Suit filed here seeks upgrading of facilities at District Jail," *The Washington Post*, March 3, 1975.

10 Karl Menninger, "The Criminal Law System," *Nebraska Law Review* 4.22–32 (1966): 27.

11 C. J. Ciaramella, "How Not To Build A Jail," *Reason* magazine, Dec. 2016.

12 Mary Ann Kuhn, "Plight of Jail Told in Court," *Metro Life*, Oct. 12, 1972, B3.

13 "The Disgrace of DC Jail," *The Washington Post*, Nov. 12, 1975.

14 Calhoun, "Remarks about a visit to the D. C. Jail," Jan. 18, 1972, URBS doc 194, Part B; Robert Pear, "Judge is chilly to city's defense concerning jail," *Washington Star*, March 5, 1975.

15 Goldfarb, "A 'Non-Architecture' Approach to Prison Reform," *The Washington Post*, Jan. 20, 1972.

16 Nixon, "Towards Freedom From Fear."

17 Paul Paulus, Garvin McCain, and Verne Cox, "A Note on the Use of Prisons as Environments for Investigating Crowding," *Bulletin of the Psychonomic Society* 1 (1973): 427–8.

18 Garvin McCain, Verne Cox, and Paul Paulus, "The Relationship between Illness Complaints and Degree of Crowding in a Prison Environment," *Environment and Behavior* 8.2 (1976): 283–290, 289.

19 Paul Paulus and Garvin McCain, "Crowding in Jails," *Basic and Applied Social Psychology* 4 (1983): 89–107, 89.

20 Garvin McCain, Verne Cox, and Paul Paulus, "The Relationship between Illness Complaints and Degree of Crowding," 288.
21 Paul Paulus, *Prison Crowding: A Psychological Perspective* (New York: Springer,1988), 6.
22 Paulus, Prison Crowding, 7.
23 Calhoun, letter to Jane Bradley, of Cohen & Rosenblum, Aug. 30, 1973.
24 John Zeisel, "Behavioral Research and Environmental Design: A Marriage of Necessity," *Design & Environment* 1 (1970): 51–66.
25 Terry Maple, "Psychology Is Alive and Well at the Zoo," CSUN online, accessed Jan. 1, 2023.
26 Maple, "Psychology Is Alive and Well at the Zoo."

CHAPTER 13: THE RAT BILL

1 Lyndon Baines Johnson, *The Vantage Point: Perspectives on the Presidency 1963–1969* (New York: Holt, Rinehart, and Winston, 1971), 84.
2 "Rat Control Rejected," in *CQ Almanac 1967*, 23rd Ed. (Washington, DC: Congressional Quarterly, 1968), 13–446. library.cqpress.com.
3 Tom Wicker, "In the Nation: Ratting on Newark," *New York Times*, July 23, 1967.
4 "U.S. Spent $141-Billion In Vietnam in 14 Years," *New York Times*, May 1, 1975, 20.
5 Martin Luther King, Jr., "Beyond Vietnam": Speech at Riverside Church Meeting, New York, N.Y., April 4, 1967, in *Eyes on the Prize: America's Civil Rights Years*, eds. Clayborne Carson et al. (New York: Penguin, 1987), 201.
6 "Still Much More To Be Done," *The New York Times*, July 23, 1967, T–131.
7 Richard Lyons, "House Kills Rat Curb Bill, Draws a Johnson Blast," *The Washington Post*, July 21, 1967.
8 Johnson, The Vantage Point, 85.
9 Drew Pearson and Jack Anderson, "The Rats Won the Debate," *The Washington Post*, July 30, 1967.
10 Paul Hope, "'68 May Be Political 'Year of Rat,'" *The Evening Star*, Aug. 7, 1967.
11 "Rats and the House," *The Washington Post*, Aug. 11, 1967.
12 Calhoun, letter to Warren Weaver, July 12, 1967, Calhoun Papers, Box 29.
13 David E. Davis, letter to Warren Weaver, July 20, 1967, Calhoun Papers, Box 29.
14 Calhoun, "A Manifesto on Rats, Rodents and Human Welfare," Draft, Aug. 2–9, 1967, Calhoun Papers, Box 15A.
15 David E. Davis, letter to Weaver, July 20, 1967, Calhoun Papers, Box 29.

CHAPTER 14: SPACE CADETS

1 Space Cadet minutes, May 6–7, 1957.
2 Space Cadet minutes, held at APA, May 28–9, 1956.
3 Space Cadet minutes, Oct. 22–24, 1959.
4 Space Cadet minutes, Oct. 11–12, 1956.
5 W.H.O., "Alcohol and Alcoholism: Report of an Expert Committee," *World Health Organization Technical Report Series* 94 (Geneva: W.H.O., 1951): 4.
6 John R. Seeley, "Alcoholism Is a Disease: Implications for Social Policy," in *Society, Culture, and Drinking Patterns*, eds. David J. Pittman and Charles R. Snyder (New York: Wiley, 1962), 593.
7 Seeley, "The Ecology of Alcoholism: A Beginning," in *Society, Culture, and Drinking Patterns,* 338, 341, 343.

8 Ian McHarg, *The House We Live In*, television series, 22 episodes, 1962–63. WCAU-TV, Philadelphia, PA.
9 Space Cadet minutes, Oct. 27, 1961.
10 Ian McHarg, *Design With Nature* (New York: The Natural History Press, 1969), 79–93, 195.
11 "West End Project Report: A Preliminary Redevelopment Study of the West End of Boston," March 1953, folder 3, box 2, Urban Redevelopment Division, Boston Housing Authority, Gans Papers, Columbia University.
12 "Charles River Park: The Wonderful Experience of Spacious In-Town Living," brochure, folder 2, box 1, Gans Papers.
13 Yiddish Book Center, "Leonard Nimoy Remembers Boston's West End Neighborhood," Wexler Oral History Project, 2014.
14 Marc Fried and Peggy Gleicher, "Some Sources of Residential Satisfaction in an Urban Slum," *Journal of the American Institute of Planners* 27 (1961): 305–315; 314, 315, 311.
15 Herbert Gans, The Urban Villagers: Group and Class in the Life of Italian Americans (New York: Free Press, 1962), xiii.
16 Space Cadet minutes, Oct. 17–18, 1957, 62.
17 Space Cadet minutes, May 26–27, 1958, 6.
18 Space Cadet minutes, Oct. 17–18, 1957, 58, 59, 59.
19 Gans, The Urban Villagers, xiv.
20 Outdoor Recreation Resources Review Commission, Conference on Leisure—Outdoor Recreation and Mental Hospital, Williamsburg, Virginia. June 1, 1961, 153.
21 Jane Jacobs, "Downtown is for People," in *The Exploding Metropolis*, ed. William H. Whyte (New York: Time, Inc., 1958), 167.
22 Jacobs, "Downtown is for People," 141.
23 William H. Whyte, "Groupthink," *Fortune* 142.146 (March 1952): 114–117; *The Organization Man* (New York: Simon and Schuster, 1956).
24 William H. Whyte, "Introduction," *The Exploding Metropolis* (New York: Time, Inc., 1958), x, 1.
25 Jacobs, "Downtown is for People," 168.
26 William H. Whyte, *City: Rediscovering the Center* (New York: Doubleday, 1988), 4.
27 William H. Whyte, *The Last Landscape* (New York: Doubleday, 1968), 337.
28 Whyte, *City*, 5, 2, 2, 7, 7.
29 McHarg, *Design With Nature*, 1–2.
30 *Multiply and Subdue the Earth*. Film. WGBH Educational Foundation (1969).
31 Outdoor Recreation Resources Review Commission, Conference on Leisure—Outdoor Recreation and Mental Hospital, Williamsburg, Virginia (June 1, 1961), 156, 158, 125–6.
32 "Comments by Edward T. Hall," Conservation Foundation Symposium, Belmont Hose, May 20-21, 1969, Calhoun Papers, Uncatalogued, Box 57.
33 Jane Jacobs, *The Death and Life of Great American Cities* (Harmondsworth: Penguin Books, 1974 [1961]), 218.
34 Calhoun, "The Ecology of Aggression: Its Relationship to Frustration and Social Withdrawal," 1962, Calhoun Papers, Box 68.
35 Calhoun, "Comments Concerning the Relationship between Mental Health and Recreation," Calhoun Papers, Box 67.
36 Outdoor Recreation Resources Review Commission, 133.
37 Space Cadet minutes, May 28–29, 1956, 168.

38 Richard Meier, "Violence: The Last Urban Epidemic," *American Behavioral Scientist*, 1968, March–April, 35–37.
39 Leonard J. Duhl, *The Urban Condition: People and Policy in the Metropolis* (New York: Simon and Schuster, 1963), xiii, ix.
40 John A. Andrew, *Lyndon Johnson and the Great Society* (Chicago: Dee, 1998), 135.
41 Duhl in Joe Flower, "Building Healthy Cities: Excerpts From A Conversation With Leonard J. Duhl," *Healthcare Forum Journal* 36.3 (1993).
42 Robert C. Wood, Whatever Possessed the President? Academic Experts and Presidential Policy, 1960–1988 (Amherst: University of Massachusetts Press, 1993), 79.
43 Duhl, "Building Healthy Cities."
44 Space Cadet minutes, March 17, 1966.
45 Johnson, "Remarks to Delegates to the National Convention, AFL-CLIO, December 12, 1967," in *Public Papers of the Presidents of the United States, Lyndon B Johnson, Book 2—July 1 to December 31, 1967* (Washington, DC: USGPO, 1968), 1125.

CHAPTER 15: VERTICAL SLUMS

1 "Slum Surgery in St. Louis," *Architectural Forum* 94 (April 1951): 128–136.
2 "Four Vast Housing Projects for St. Louis: Hellmuth, Obata and Kassabaum, Inc.," *Architectural Record* 120 (Aug. 1956): 182–189.
3 "The Writing on the Wall," *Horizon*, television documentary, BBC (Feb. 11, 1974).
4 "City news," *JOH*, 2 (1972): 77.
5 Ada Louise Huxtable, "A Prescription for Disaster," *The New York Times*, Nov. 5, 1972: D–23.
6 Charles Jencks, The New Paradigm in Architecture: The Language of Post-Modernism (New Haven: Yale UP, 2002), 9.
7 Huxtable, "A Prescription for Disaster," D–23.
8 A. R. Gillis, "Strangers Next Door: An Analysis of Density, Diversity, and Scale in Public Housing Projects," *Canadian Journal of Sociology* 8 (1983): 1–20, 3–4.
9 Space Cadet minutes, May 26–27, 1958, 31.
10 Space Cadet minutes, May 6–7, 1957, 120.
11 Izumi, "Memorandum to Supplement Privacy Research Project" (Dec. 13, 1967), Kansas, Box 58, 2595.
12 Qtd. in William Kloman, "E.T. Hall and the Human Space Bubble," *Horizon*. 4.4 (1967): 42–47; 46.
13 Hall, *Hidden Dimension*, 169.
14 Mildred and Edward Hall, "Pruitt Igoe Action Team, Re: The Relationship Between Design and Tenant Mix" (Nov. 15, 1971), Hall Papers, Arizona, Box 13, Folder 24.
15 Hall to Donald Henderson, University of Pittsburgh, Aug. 20, 1971, Hall Papers, Arizona, Box 13, Folder 25.
16 Lee Rainwater, "Fear and the House-as-Haven in the Lower Class," *Journal of the American Institute of Planners* 32.1 (1966): 23–31, 29.
17 "Housing without fear," *Time*, Nov. 27, 1972, 69–70, 69.
18 Huxtable, "A Prescription for Disaster," D–23.
19 Oscar Newman, *Defensible Space: Crime Prevention Through Urban Design* (New York: Macmillan, 1972), 107, 2.
20 Calhoun, "Some Professional Activities of John B. Calhoun," Calhoun Papers, Box 1, Folder 79.
21 Space Cadet minutes, Oct. 22, 23, 24, 1959, p. 93.

22 Hall, "Public Housing for Low Income families," Hall Papers, Arizona, Folder 24.
23 Hall to Donald Henderson, Aug. 20, 1971, Hall Papers, Arizona, Box 13, Folder 25.
24 L. M. Friedman, *Government and Slum Housing: A Century of Frustration* (Chicago: Rand McNally and Co, 1968), 145.
25 Duhl in Flower, "Building Healthy Cities."
26 Office of the White House Press Secretary, Sept. 19, 1973, Oscar Newman Papers, Columbia University, Box 1.
27 Oscar Newman, letter to Phillip Herrera, Oct. 6, 1972, Newman Papers, Box 1.
28 J. R. Knoblauch, "Going Soft: Architecture and the Human Sciences in Search of Institutional Forms" (PhD Princeton, 2012), 6.
29 William Russell Ellis, "Meticulous, Through and Ideologically Corrupt," *The Daily Californian*, 1973, 3.
30 Duhl, letter to Calhoun, March 3, 1977, Calhoun Papers Box 87.
1 C. B. Huffaker, "Experimental Studies on Predation: Dispersion Factors and Predator-Prey Oscillations," *Hilgardia* 27.14 (1958): 343–383; 344.

CHAPTER 16: POOLESVILLE

1 Jimmy Sonni and Rob Goodman, *A Mind At Play: The Brilliant Life of Claude Shannon* (New York: Simon Schuster, 2017), 141.
2 Calhoun, "Antiquaria," *317 P.H.*, unfinished novel (1963), ch. 7, 5.
3 Minutes, March 11, 1966, visit of NIMH Councilors to the NIH Animal Center at Poolesville, Maryland, Calhoun Papers, Box 22.
4 Calhoun, letter to Simone Swan, The Menil Foundation, Houston, Texas, Jan. 22, 1975, Calhoun Papers, Box 101.
5 Calhoun, "Some Definitions Relating to Behavioral Systems," Jan. 30, 1972, Calhoun Papers, Box 101.
6 Steven Shapin, "Paradigms Gone Wild," *London Review of Books* 45.7 (2023).
7 Calhoun, "Notes and Quotes which May Serve to Promote Dialogue Related to URBS' Aims and Activities," Calhoun Papers, Box 1B.
8 Calhoun, R_x*evolution*: 1–2.
9 Stanley, letter to Calhoun, Jan. 12, 1966, Calhoun Papers, Box 22.
10 Paul D. MacLean, letter to Livingston, Nov. 14, 1958, MacLean Papers, NLM, Box 6, Folder 40.
11 MacLean, "Psychosomatic Disease and the Visceral Brain. Recent Developments Bearing on the Papez Theory of Emotion," *Psychosomatic Medicine* 11 (1949): 338–353, 351.
12 Claudio Pogliano, "Lucky Triune Brain. Chronicles of Paul D. MacLean's Neuro-Catchword," *Nuncius* 32 (2017): 330–75.
13 MacLean, *A Triune Concept of the Brain and Behaviour* (Toronto: University of Toronto Press, 1973), 7.
14 MacLean, "Meeting with Dr. Livingston and Dr. Jack Calhoun on May 11, 1959, regarding the behavioral farm," qtd. in Pogliano, "Lucky Triune Brain."
15 MacLean, *Triune Concept of the Brain*, 59.
16 John F. Kennedy, Economic Club Dinner, Chicago, IL, Oct 9, 1957.
17 Heinz von Foerster, Patricia M. Mora, and Lawrence W. Amiot, "Doomsday: Friday, 13 November, A.D. 2026." *Science* 132.3436 (1960): 1291–1295; 1295.
18 Fairfield Osborn, *Our Crowded Planet: Essays on the Pressures of Population* (London: Allen and Unwin, 1963), 9.

19 US Government. "Population Crisis," Hearings before the Permanent Subcommittee on Foreign Aid Expenditures of the Committee on Government Operations, United States Senate, Eighty-Ninth Congress, Second Session. Washington: US Government Printing Office; exhibits 100, 101 (March 31, 1966), 631, 638, 647; (April 6, 1966), 737, 719.
20 Paul R. Ehrlich, *The Population Bomb* (New York: Ballantine, 1968), 66–67, 161, 165-66.
21 Nathan Keyfitz, "United States and World Populations," in *Resources and Man* (San Francisco: W.H. Freeman and Company, 1969), ch. 3.
22 Calhoun, "The Revolution of Compassion," Oct. 18, 1969, Calhoun Papers, Box 101.
23 Calhoun, "Space and the Strategy of Life," in *Behavior and Environment: The Use of Space by Animals and Man*, ed. Aristide H. Esser (New York: Plenum, 1971), 372, 379.
24 MacLean, *Triune Concept of the Brain*, 7.

CHAPTER 17: UNIVERSE 25 / THE KESSLER PHENOMENON

1 Calhoun, letter to Harold Fishbein, Sept. 12, 1964.
2 Calhoun, "Space and the Strategy of Life," 331.
3 Calhoun, "Ecological Factors in the Development of Behavioral Anomalies," in *Comparative Psychopathology: Animal and Human*, eds. Joseph Zubin and Howard F. Hunt (New York: Grune and Stratton, 1967), 46.
4 Calhoun, "Position Opening," Calhoun Papers, Box 96.
5 Calhoun, R_x*evolution*, 1–3.
6 Calhoun, "What Sort Of Box?" 23.
7 Calhoun, R_x*evolution*, 1–3.
8 Maya Pines, "How the Social Organization of Animal Communities Can Lead to a Population Crisis Which Destroys Them," *NIMH Program Reports, no. 5*, ed. Julius Segal (1971), 158–173.
9 Calhoun, "What Sort Of Box?" 21.
10 Calhoun, R_x*evolution*, 1–2, 1–18a, 1–15, 1–12, 1–5, 1–5, 1–17, 1–9.
11 Calhoun, "What Sort Of Box?" 23.
12 Calhoun, R_x*evolution*, 1–21, 1–22.
13 Calhoun, John B., "Death Squared: The Explosive Growth and Demise of a Mouse Population," *Proceedings of the Royal Society of Medicine* 66 (Jan. 1973), 80–88; 85, 85.
14 Calhoun, R_x*evolution*, 1–25.
15 Calhoun, "Universal Autism: Extinction Resulting from Failure to Develop Social Relationships," Draft manuscript from presentation at Georgetown University Family Center (1986), URBS Doc 275A.
16 Calhoun, "What Sort Of Box?" 23, 24.
17 Calhoun, R_x*evolution*, 1–26.
18 Calhoun, "Universal Autism," 8.
19 Tom Huth, "Ten Boxes of Dead Mice Could Be Us," *The Washington Post*, Feb. 8, 1973.

CHAPTER 18: POPULARITY CONTROL

1 Stewart Alsop, "Dr. Calhoun's Horrible Mousery," *Newsweek* (Aug. 17, 1970), 96.
2 Calhoun, R_x*evolution*, 1–29.
3 "Crowding Produces Weirdo Rodents," *Rocky Mountain News*, Feb, 11, 1970.

4 Frank Sartwell, "The Small Satanic Worlds of John B. Calhoun," *Smithsonian Magazine*, April 1970.
5 Robert Burdick, "Ecologist Views Mice, Men," *Kansas City Times*, Oct. 7, 1970.
6 Pines, "How the Social Organization . . . ," 158–173.
7 Nina Laserson, "It's Not Every Day You Walk into a Laboratory Whose Mission Is to Save the World," *Innovation* 23 (1971): 15–25, 25.
8 Matthew Wisnioski, "*Innovation* Magazine and the Birth of a Buzzword," *IEEE Spectrum*, online (Jan. 29, 2015).
9 Tom Huth, "Ten Boxes of Dead Mice Could Be Us."
10 F. Fraser Darling and Raymond F. Dasmann, *A Conversation on Population, Environment, and Human Well-Being* (Washington, DC: The Conversation Foundation, 1971), 66.
11 Mary Steichen Calderone, "Human Cost Accounting," in *The Complete Book of Birth Control* (North Hollywood, Calif.: American Art Agency, 1965), 9.
12 "People Pollution," *Medical World News*, Dec. 19, 1969.
13 Walter E. Howard, "The Population Crisis Is Here *Now*," *BioScience* 19 (1969): 779–784; 784, 779.
14 Roy O. Greep, "Prevalence of People," *Perspectives in Biology and Medicine* 12 (1969): 332–343; 340, 338.
15 Calhoun, letter to Lee Rainwater, Jan. 6, 1966, Calhoun Papers, Box 4A.
16 Calhoun, letter to Eberhart, Dec. 12, 1972, Calhoun Papers, Box 101.
17 Calhoun, "$R_x$evolution, Tribalism, and the Cheshire Cat: Three paths from Now," *Technological Forecasting and Social Change*, 4 (1973): 263–282, 274, 275.
18 Calhoun, Ecology and Sociology of the Norway Rat, 216.
19 Calhoun,"What Sort Of Box?" 29–30.
20 Calhoun, *$R_x$evolution*, 2–11.
21 Calhoun, "Death Squared," 80, 80, 88.
22 Calhoun, *$R_x$evolution*, 2–11, 2–11
23 Hall, letter to John J. Christian, April 20, 1962.

CHAPTER 19: A PRESCRIPTION FOR EVOLUTION

1 C. Henry Kempe, Frederic N. Silverman, Brandt F. Steele, William Droegemueller, and Henry K. Silver, "The Battered-Child Syndrome," *Journal of the American Medical Association* 181 (1962): 17–24.
2 Calhoun, *$R_x$evolution*, 4–1.
3 Calhoun, "$R_x$evolution, Tribalism, and the Cheshire Cat," 270.
4 Calhoun, letter to Eberhart and MacLean, "Partial Sabbatical, June 1, 1974 to December 31, 1974," Calhoun Papers, Box 2B.
5 Sabbatical log, June 4, 1974. Calhoun Papers, Box 101.
6 Sabbatical log, Sept. 10, 1974, Calhoun Papers, Box 101.
7 Carol Houck Smith, Norton, letter to Calhoun, Aug. 17, 1970, Calhoun Papers, Box 5A.
8 Calhoun, letter to Crook, Nov. 25, 1976, Calhoun Papers, Box 1.
9 Henry Scarupa, "Of Mice and Men and Escaping the Ultimate Pathology," *Baltimore Sun Magazine* (June 13, 1976).
10 Calhoun, "Scientific Quest for a Path to the Future," *Populi* 3.1 (1976): 19–28, 20, 23.
11 R. B. Lockard, "The Albino Rat: A Defensible Choice or a Bad Habit?" *American Psychologist* 23 (1968): 734–42, 739.

12 R. Robinson, *Genetics of the Norway Rat* (Oxford: Pergamon Press, 1965), 514.
13 Calhoun, letter to Milton Rubin, Sept. 3, 1964, Calhoun Papers, Box 1.
14 Calhoun, "Scientific Quest," 25, 24.
15 Calhoun, "Experimental Development of Tribal Affinities in Rats," Dec. 18, 1960, Calhoun Papers, Box 99.
16 Calhoun, "Scientific Quest," 26.
17 Scarupa, "Of Mice and Men and Escaping the Ultimate Pathology."
18 Wray Herbert, "The (Real) Secret of NIMH," *Science News* 122 (1982): 92–93.
19 Calhoun, "Scientific Quest," 25.
20 Calhoun, letter to Garfield, Dec. 9, 1974, Calhoun Papers, Box 2B.
21 Diary, Sept. 10, 1974, Calhoun Papers, Box 101.
22 Calhoun, "Seven Steps from Loneliness," in *The Anatomy of Loneliness*, eds. Joseph Hartog, J. Ralph Audy, and Yehudi A. Cohen (New York: International Universities Press, 1981), 132, 132, 133.
23 Calhoun, letter to Garfield, Dec. 9, 1974, Calhoun Papers, Box 2B.
24 Calhoun, "Seven Steps from Loneliness," 132, 133.
25 Calhoun, "What Sort Of Box?" 30.
26 Calhoun, R_x*evolution*, 4–2.

CHAPTER 20: SYSTEM FAILURE

1 Committee on Co-Administration of Service and Research Programs of the National Institutes of Health, Institute of Medicine, *Research and Service Programs in the PHS: Challenges in Organization* (Washington: National Academy Press, 1991), 30, 34.
2 Calhoun, Memorandum, May 5, 1965.
3 Calhoun, letter to President Nixon, Jan. 3, 1974: Calhoun Papers, Box 28.
4 Laserson, "It's Not Every Day…" 17.
5 Calhoun, letter to Simone Swan, The Menil Foundation, Jan. 22, 1975, Calhoun Papers, Box 101.
6 "Rats Given 'Culture' As Defense Against Overcrowding," *ADAMHA News*, Sept. 22, 1978, 3, 3.
7 Calhoun, Annual report summary 1979, URBS, LBEB, NIHAC, June 25, 1979, Calhoun Papers, Box 98.
8 Calhoun, Annual review, Laboratory of Brain Evolution and Behavior, October 1, 1982, to August 20, 1983, Calhoun Papers, Box 102.
9 H. G. Wells, *World Brain* (Garden City, NY: Country Life Press, 1938), 20, 21, 33, v.
10 Calhoun, ed., *Environment and Population: Problems of Adaptation: An Experimental Book Integrating Statement by 162 Contributors* (New York: Praeger, 1983), ix, ix, vii.
11 Gerald L. Young, "Review of *Environment and Population*, by John B. Calhoun," *Human Ecology* 12.4 (1984): 499–507, 501, 506.

CHAPTER 21: ECCO LIBRIUM

1 Thomas A. Ban, "The Role of Serendipity In Drug Discovery," *Dialogues in Clinical Neuroscience* 8.3 (2006): 335–344; Randall Lowell, "This Week's Citation Classic," *Current Therapeutic Research* 47 (1980): 337.
2 Joseph V. Brady, "Conversation with Joseph V. Brady," *Society for the Study of Addiction*. 100 (2005): 1805–1812, 1807.
3 Smith, Kline, and French, "Thorazine," Advertisement (1955).

4 Jonathan Cole, qtd. in "Tranquilizers—Appendix." *Hearings before the Subcommittee on Antitrust and Monopoly*. 86th Congress. 2nd Session. 11.17 (1960): 9453.
5 L. A. Wikler, "The Use of Chlropromazine as an Anti-emetic in Children," *Archives of Pediatrics* (NY) 72.6 (1955): 197–199.
6 Space Cadet minutes, May 6–7, 1957, 166–168.
7 Ardrey, letter to MacLean, Aug. 6, 1972, Paul MacLean Papers, NLM, Box 1.
8 Calhoun, "Annual Report: Conceptual Adaptation and Evolution—July 1, 1978, through September 30, 1979," Calhoun Papers, Box 84.
9 Claude S. Fischer, "Sociological Comments on Psychological Approaches to Urban Life," in *Advances in Environmental Psychology, vol. 1, The Urban Environment*, ed. A. Baum, J. M. Singer, and S. Valins (Hillsdale, N.J.: Erlbaum, 1978), 132.
10 Fischer and Baldassare, "How Far from the Madding Crowd?" *New Society* 32 (1975): 531–533, 531.
11 E.V. Walter, "Dreadful Enclosures: Detoxifying and Urban Myth," *European Journal of Sociology* 18.1 (1977): 150–159, 155, 155.
12 Robert C. O'Brien, *Mrs Frisby and the Rats of NIMH* (New York: Athenuem, 1971), 104, 160.
13 Wray Herbert, "The (Real) Secret of NIMH," *Science News* 122 (1982): 92–93.
14 Sandy Rovner, "Rats! The Real Secret of NIMH," *The Washington Post*, July 21, 1982, B4, 1, 4, 1.
15 Calhoun, Annual Report of the Unit for Research on Behavioral Systems, Sept. 30, 1982, Calhoun Papers, Box 106.
16 Matthew Dumont, DHEW, NIH, Meeting on Primary Prevention, Chevy Case, MD, June 27, 1969, Calhoun Papers, Box 77, 234.
17 Calhoun, Annual Report Summary, URBS, LBEB, NIHAC, June 25, 1979. Calhoun Papers, Box 98.
18 Herbert, "The (Real) Secret of NIMH," 92.
19 Rovner, "Rats! The Real Secret of NIMH," 4.
20 Frederick K. Goodwin, Memorandum to William E. Mayer (March 9, 1983), in Calhoun, "A 'Hitchhiker's Guide' to Three Worlds" (Aug. 1986), Calhoun Papers, Box 18.
21 Goodwin, "Current and Future Plans," NIMH (Sept. 22, 1983), Calhoun Papers, Box 95.
22 William T. McKinney Jr., Laurens D. Young, Stephen J. Suomi, John M. Davis, "Chlorpromazine Treatment of Disturbed Monkeys," *Arch Gen Psychiatry* 29.4 (1973): 490–494; Stephen J. Suomi, Stephen F. Seaman, Jonathan K. Lewis, Roberta D. DeLizio, William T. McKinney, "Effects of Imipramine Treatment of Separation-Induced Social Disorders in Rhesus Monkeys," *Arch Gen Psychiatry* 35.3 (1978): 321–325; Stal Saurav Shrestha, et al., "Fluoxetine Administered to Juvenile Monkeys: Effects on the Serotonin Transporter and Behavior," *The American Journal of Psychiatry* 171.3 (2014): 323–331.
23 Stephen J. Suomi, "Maternal Behavior by Socially Incompetent Monkeys: Neglect and Abuse of Offspring," *Journal of Pediatric Psychology* 3 (1978): 28–34; 28, 29.
24 Goodwin, "Current and Future Plans."
25 Ronald Reagan, Memorandum for Heads of Executive Departments and Agencies (Dec. 21, 1983).
26 Calhoun, letter to Chief of Clinical Science, IRP, NIMH, Jan. 30, 1986, Calhoun Papers, Box 98.

27 L. C. Kolb, S. H. Frazier, P. Sirovatka, "The National Institute of Mental Health: Its Influence on Psychiatry and the Nation's Mental Health," in *American Psychiatry after World War II: 1944–1994*, eds. R. W. Menninger, J. C., Nemiah (Washington, DC: American Psychiatric Press, 2000), 207–31, 223; Allan V. Horwitz, "How an Age of Anxiety Became an Age of Depression," *Milbank Quarterly* 88 (2010): 112-38, 121.
28 Calhoun, letter to Dominque de Menil, April 15, 1986, Calhoun Papers, Box 1.
29 Calhoun, letter to John Herbers, Sept. 4, 1986, Calhoun Papers, Box 18.
30 Mayer, qtd. in Calhoun, letter to Dominque de Menil, April 15, 1986.
31 Calhoun, "A 'Hitchhiker's Guide' to Three Worlds," n.p.

CODA: QUIETUS

1 Calhoun, *The Migrant* 6.3 (1935): 71.

ACKNOWLEDGMENTS

We began writing about the rats of NIMH while working at the London School of Economics in 2007, and always planned to compile a fuller account of Jack Calhoun's life and work, but the book never would have come to pass if it had not been for Athena Bryan, who persuaded us to write it and Melville House to publish it. We're very grateful to her, and to our editorial team, Carl Bromley, Michelle Capone, and Amber Qureshi.

There have been many friends and colleagues over the years who have provided support and sound advice. We would like to thank the members of the "Facts" project at LSE, particularly Mary Morgan and Sabina Leonelli; and at the University of Exeter, Mark Jackson and Matt Smith, and thanks also to David Cantor, Rhodri Hayward, Rob Kirk, Andrew Mendelsohn, Josh Ramsden, Mike Sappol, Patrick Sommerville, and Duncan Wilson, for their help and advice over the years. At various points, Chip Adams, James Bruhn, Craig Fox, David Boyd Haycock, David Mitchell, Gary Wingfield, and Jonathan Wilson all offered helpful commentary and counsel.

The National Library of Medicine in Bethesda, particularly John Rees, who was invaluable for exploring the Calhoun papers, which were uncatalogued and held off-site for much of our research for the book; the Chesney Medical Archives at Johns Hopkins, particularly Andy Harrison, for helping to track down material relating to Curt Richter and the Rodent Ecology Project; as well as archivists and librarians at the Rockefeller Archive Center in Tarrytown, New York, the Kenneth Spencer Research Library at the University of Kansas, Columbia University Archives, the University of

Arizona, and the American Heritage Center, University of Wyoming which originally held many of the Calhoun papers before their transfer to the NLM. The resources made available through archive.org were invaluable, and we urge you to donate.

Our deep gratitude to the family of John B. Calhoun, particularly Edith and Cat Calhoun, who were very generous in the provision of information as well as sharing photographs and stories of Jack's life.

Finally, and most importantly, we would like to thank our families for their patience and support: Emma, Sam, Lily, Sylvie, and Frederick; Amy and Chris, Di and Alan, Jo, Jenny; and Charlotte, Martha, and Caleb.

INDEX

NOTE: PAGE NUMBERS IN ITALICS REPRESENT IMAGES AND PHOTOGRAPHS.